STUDENT UNIT GUIDE

NEW EDITION

WJEC AS Geography Unit G2
Changing Human Environments

David Burtenshaw and Sue Warn

PHILIP ALLAN

Philip Allan Updates, an imprint of Hodder Education, an Hachette UK company, Market Place, Deddington, Oxfordshire OX15 0SE

Orders

Bookpoint Ltd, 130 Milton Park, Abingdon, Oxfordshire OX14 4SB
tel: 01235 827827
fax: 01235 400401
e-mail: education@bookpoint.co.uk
Lines are open 9.00 a.m.–5.00 p.m., Monday to Saturday, with a 24-hour message answering service.
You can also order through the Philip Allan Updates website: www.philipallan.co.uk

© David Burtenshaw and Sue Warn 2012

ISBN 978-1-4441-6200-4

First printed 2012
Impression number 5 4 3 2
Year 2016 2015 2014 2013 2012

Cover photo: Aania/Fotolia

Printed in Dubai

Hachette UK's policy is to use papers that are natural, renewable and recyclable products and made from wood grown in sustainable forests. The logging and manufacturing processes are expected to conform to the environmental regulations of the country of origin.

This material has been endorsed by WJEC and offers high quality support for WJEC qualifications. While this material has been through a WJEC quality assurance process, all responsibility for the content remains with the publisher.

P01996

WJEC examination questions used with permission of WJEC.

Contents

Content Guidance

Questions & Answers

Getting the most from this book

Questions & Answers

Exam-style questions

Examiner comments on the questions

Tips on what you need to do to gain full marks, indicated by the icon **e**.

Sample student answers

Practise the questions, then look at the student answers that follow each set of questions.

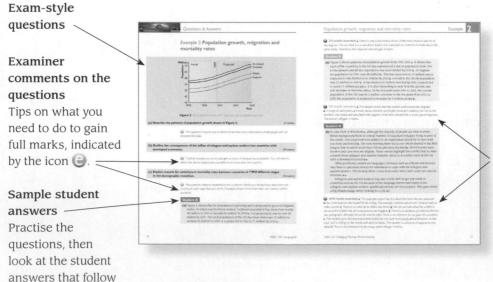

Examiner commentary on sample student answers

Find out how many marks each answer would be awarded in the exam and then read the examiner comments (preceded by the icon **e**) following each student answer. Annotations that link back to points made in the student answers show exactly how and where marks are gained or lost.

About this book

The purpose of this guide is to help you understand what is required to do well in **Unit G2: Changing Human Environments**. The specification is available on the WJEC website: **www.wjec.co.uk**. This paper may be taken in Welsh. The comments all apply no matter what your language medium.

This guide is divided into two sections. The **Content Guidance** section gives you an introduction to the two themes that you have to study. It contains several useful diagrams and maps, which you may be able to use in the examination. Unit G2 is an assessment which relies on you writing mini essays in continuous prose. To help you develop your geographical writing skills the **Questions & Answers** section provides guidance on how to approach the unit test and includes examples of the types of question that you will see in the examination.

Content Guidance

Theme 1: Investigating population change

The emphasis here is on **change** and therefore the concepts, ideas, issues and problems will all reflect the theme of change and adapting to change. Population is dynamic for it is always changing as people are born, die and migrate: it is the flow of people through time.

1.1 What is demographic change?

Population growth and changing population are among our greatest challenges especially as they are occurring in those countries least equipped to cope. The seven billionth person was born in 2011. **Systems theory** provides a framework for studying population because it enables us to see any population as dynamic. **Demography** is the study of population's vital statistics. This section looks at the tools you need to demonstrate your understanding of the fundamental principles. Population change can be measured at a variety of scales. The world's population doubled between 1960 and 2000. However, the rate of increase varies. The UK's population will take 433 years to double compared with 27 years in Africa. Since 1950 Africa and Asia have contributed 81% of the world's population growth whereas Europe has contributed only 4.4%. As a result, the distribution of the world's population is changing. Do bear in mind that the art of forecasting is very imprecise and is based mainly on projecting existing trends.

There are now 24 countries with over 50 million people (there were 9 in 1950). In 2010, China had approximately 1338.1 million and India 1188.8 million. The countries with the highest increases in 2000 were Islamic. By 2050 it is projected that the population of Yemen will rise by 121%, Saudi Arabia by 70% and Pakistan by 81%.

On 1 January 2011 the estimated population of the UK was 62 million. Key patterns that you need to be aware of are **density** and **change** (natural and migration). In the UK population is growing at 0.6% per year compared with 0.3% in the 1990s. Women are having more babies but there has also been a 20% rise in women of child-bearing age as a result of migration. Migration has contributed less to the absolute growth in numbers because many migrants return home and this is combined with the emigration of UK citizens.

Knowledge check 1

Why is population change a challenge?

Population and demographic data

Population data are normally collected by ten-yearly national censuses. These give the absolute numbers at that time, either where people are, or where they normally are. Some data are sampled; normally 10% of households is the sample size. Censuses may lack accuracy because people do not complete forms. These days sampling between censuses by national statistical offices and by organisations such as the UN does give us more detail for intercensal periods— see Tables 1 and 2.

Key terms

Natural increase is the excess of births over deaths. **Crude birth rate** is the ratio of the number of births per annum to the total population expressed per 1000 people. **Total fertility rate** is measured as the number of births to a woman over her lifetime. This can be refined to examine the number of births to a specified age group per 1000 women. In MEDCs this rate needs to be 2.1 children per woman aged 15–45 to maintain the level of population. In Italy the rate was 1.3 in 2005 while in the UK it was 1.7. In Central African states in 2005 the average rate was 5.6 and in Islamic countries it was 4.1. Birth rates and fertility rates can be influenced by:

- culture, for example religious beliefs and dogma
- political pressures to increase or decrease the population
- social norms, for example later marriage
- events such as the end of a war

Crude death rate is the ratio of the number of deaths per annum to the total population, expressed per 1000 people. The **age-specific mortality rate** refines this. Hygiene, diet and medical advances all affect the rise and fall of these rates. A key rate is the **infant mortality rate**, which measures the number of deaths of children under 1 year old per 1000 live births. The **maternal mortality rate** measures the deaths of women in childbirth per 100,000 live births. These last two measures are frequently used to measure the development of a country.

Life expectancy is the average number of years a person is expected to live. It can be refined by age groups and gender. It was 80 for children born in the UK in 2010 (MEDC average 77) whereas in Lesotho it was 41 (LEDC average 67) and for all Southern African countries it was 55.

Migration is the movement of people from one administrative area to another, which results in the change of permanent residence. **Migration balance** is the excess of in-migration over out-migration, or vice versa. **In-migration** is the flow of people over a given time period (year or decade) into a country/area, on either a temporary or a permanent basis. **Out-migration** is the opposite flow.

Net population change is the change in a country over a period of time when both natural change and migration change have been taken into account. It is normally expressed as a number or a percentage. Figure 1 shows data for 1995–2000.

Knowledge check 2

When was the last census in the UK?

Examiner tip

You will gain credit if you can define the terms accurately and provide an example.

Knowledge check 3

What are the crude birth and death rates?

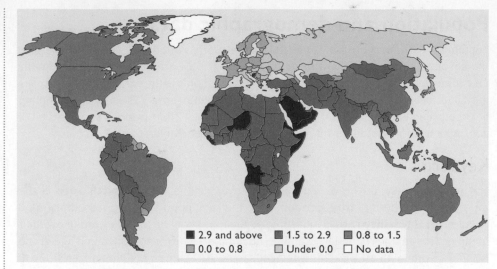

Figure 1 Annual average rate of population change, 1995–2000 (%)

Displaying demographic and population data

Geographers and demographers portray population in a variety of ways. Each of the measures above can be mapped at any scale from the global to the local (city).

Density maps show the number of people per unit of land area (persons per hectare). This is shown as a **choropleth map** (shading or colouring areas of different densities). This technique is used to illustrate most of the data above. Density maps ignore the ability of the land/country to support the population living within its boundaries. **Physiological density** measures the number of people supported per hectare of arable land, thus measuring a country's ability to support itself from within its own boundaries.

Distribution may be portrayed by dots or proportional symbols. It is relatively unsatisfactory because symbols overlap, but it can prove useful at a local scale where there is more room to portray data.

Examiner tip

Be able to discuss the strengths and weaknesses of all types of map. The strengths and weaknesses of density, choropleth, topological and flow line maps are the same no matter what data they portray.

Knowledge check 4

What does Figure 2 tell you about the distribution of the world's population?

Figure 2 Topological map of the world's population

Topological maps portray the global population challenge more graphically. The area of each country (or administrative area) is in proportion to its total population or other total being measured (Figure 2).

Migration may be portrayed using choropleth and dot map techniques but it is best portrayed on **flow maps** showing the direction and scale of movements by the width of the flow line.

The **age–sex pyramid** classically portrays gender and age. Pyramids are an indicator of development. They are used to study the demographic characteristics of individual groups in an area. Figures 3a, b and c show some pyramids for the residents of Singapore. In order to make pyramids comparable, all data should be expressed in percentages of the total population. However, many sources use absolute numbers, which make comparisons more difficult.

Examiner tip

When given data on a world map always try to name countries to support your answer. It pays to have a world political map to consult when looking at world population data maps.

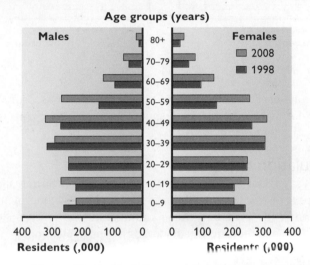

Figure 3a Population age pyramid for Singapore

Knowledge check 5

Describe the pyramids for the population of Singapore. What age groups have the greatest proportions in 2008? What are the differences between the Chinese and Malay populations in 2000?

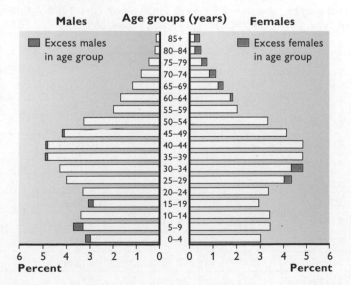

Figure 3b Chinese in Singapore (2000) as % of Chinese population (2.5 million)

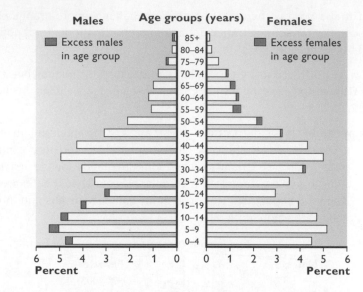

Figure 3c Malays in Singapore (2000) as % of Malay population (0.45 million)

Graphs can also be used to show trends. **Line graphs** are used to show changed rates of growth for many of the demographic indicators. **Bar graphs** are used to illustrate time series for a set of places.

Global population change

Table 1 provides a range of indicators for selected countries. The data illustrate issues such as the impact of war (Afghanistan) on life expectancy. Improving health care and its impact on death rates, birth rates and infantile mortality is clear. However, there are exceptions and they are worth looking at and explaining.

Table 1 Population data for selected countries, 2010 (1999 in brackets)

Country	Population (in millions)	Life expectancy at birth (years)	Crude death rate (per 1000)	Crude birth rate (per 1000)	Infant mortality rate (per 1000 live births)	% population 65+
Afghanistan	29.1 (22.8)	44 (46)	18 (22)	39 (50)	155 (n/a)	2
Vietnam	88.9 (78.7)	74 (68)	5 (6)	17 (20)	15 (37)	8
India	1188.8 (998)	64 (63)	7 (9)	23 (26)	53 (71)	5
China	1338.1 (1253.6)	74 (70)	7 (7)	12 (16)	21 (30)	8
Bangladesh	164.4 (127.7)	66 (58)	7 (9)	22 (28)	45 (61)	4
Chad	11.5 (7.5)	49 (47)	17 (16)	46 (45)	130 (101)	3
Sierra Leone	5.8 (4.9)	47 (38)	16 (25)	40 (45)	89 (168)	2
Burkina Faso	16.2 (11.6)	53 (71)	12 (19)	46 (44)	81 (105)	3
Burundi	8.5 (6.7)	50 (43)	15 (20)	36 (41)	98 (105)	3
Rwanda	10.4 (8.3)	51 (41)	14 (22)	42 (45)	102 (123)	2
Egypt	80.4 (67.2)	72 (67)	6 (7)	27 (26)	28 (47)	4
Uganda	33.8 (21.5)	52 (40)	13 (19)	47 (46)	76 (88)	3
Malaysia	28.5 (22.7)	74 (72)	5 (4)	21 (24)	9 (8)	5

Country	Population (in millions)	Life expectancy at birth (years)	Crude death rate (per 1000)	Crude birth rate (per 1000)	Infant mortality rate (per 1000 live births)	% population 65+
Singapore	5.1 (4)	81 (77)	4 (5)	10 (13)	2.2 (3)	9
Thailand	68.1 (60.8)	69 (69)	8 (7)	15 (17)	44 (28)	7
Mexico	110.6 (97.3)	76 (73)	5 (5)	19 (27)	17 (29)	6
Brazil	193.3 (167.9)	73 (67)	6 (7)	17 (17)	24 (32)	7
Russian Federation	141.9 (147.1)	68 (67)	14 (14)	12 (9)	8.2 (16)	13
Bulgaria	7.5 (8.2)	72 (71)	14 (14)	11 (8)	9 (14)	18
Germany	81.6 (82.1)	80 (77)	10 (10)	8 (9)	3.5 (5)	20
France	63 (58.8)	81 (78)	9 (9)	13 (13)	3.6 (5)	17
Italy	60.5 (57.6)	82 (78)	10 (10)	10 (9)	3.6 (5)	20
Japan	127.4 (126.6)	83 (80)	9 (8)	9 (10)	2.6 (4)	23
USA	309.6 (278.2)	78 (77)	8 (9)	14 (15)	6.4 (7)	13
UK	62.2 (59.8)	80 (77)	9 (11)	13 (12)	4.7 (6)	16

(Source: World Population Data Sheet 1999 and 2010)

Knowledge check 6

Try explaining the differences in infant mortality rate in Table 1.

Health
- Control of disease — death control
- Birth control — contraceptive usage
- Infant mortality
- Diet — malnutrition
- Doctors, nurses
- Sexual health
- Sanitation

Welfare
- Levels of care for the elderly
- Improving living standards
- Availability of radio and the media
- Clean water supply

Education
- Health education
- Age of ending compulsory schooling
- Female years in education
- Levels of tertiary education
- Literacy levels

Cultural
- Religious attitude to birth control
- Status gain from having children
- Role of women in society
- Role of extended family
- Sexual morality

Political
- Taxation to support services
- Economy and resources to support numbers
- Impact of war and conflict
- Facilitating access to health care and contraception

Environmental
- Frequency of major hazards
- Environmental conditions that breed killer diseases

Stage of development
- Use the Rostow model as a basis
- Social developments, such as women's rights
- Wealth from resources

Figure 4 Explaining global variations in demographic indicators

More important than the actual distributions and figures are the reasons for them. Figure 4 illustrates the key factors explaining global variations in demographic indicators. The main factor is the **stage of development.** This in turn is the product of the factors shown in the diagram. Remember, the indicators of demographic health and all of the data are fixed in time. The reality is that countries and regions are all dynamic and changing. Countries develop both economically and socially, and this affects population change over time.

1.2 How and why do populations change naturally?

The most common explanation for change is the **demographic transition** (Figure 5). Countries have distinctive characteristics at each stage. No country is now in Stage 1 although there may be peoples within some countries who are still at Stage 1. A key point is that the model ignores migration which is significant for many countries.

Figure 5 Five-stage demographic transition model

Figure 6 Cape Verde birth and death rates related to the demographic transition, 1930–2000

Stage 2 (e.g. Cape Verde; Figure 6)

- Population 500,000 in 2010.
- Accelerating growth with a large gap between birth rate (25/000) and death rate (6/000) in 2010.
- Medical revolution enabling death control.
- Lowering infant mortality (25/000).
- Increasing urbanisation and industrialisation — 60% urban.

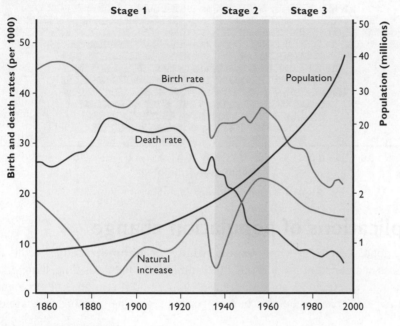

Figure 7 Chile birth and death rates related to stages in the demographic transition, 1860–2000

Stage 3 (e.g. Chile; Figure 7)

- Population growth because birth rate has fallen
- Social customs have changed.
- Greater materialism and altruism — concern for life chances.
- Economic changes have encouraged fewer children to be born.
- Economy diversifies into services.
- Chile is approaching Stage 4.

Stage 4 (e.g. Denmark)

- Low growth.
- More women in the labour force.
- Lifestyle changes — higher incomes, more leisure and increased birth control.
- Cohabitation and delayed parenthood.
- Immigration prevents total decline.
- Entering stage 5.

Stage 5 — the second demographic transition (e.g. Germany; Figure 8)

- Very low birth rate and low numbers of children per woman.
- MEDCs with increasingly old-age populations, immigration to fill gaps in workforce and higher emigration of young people and highly skilled to the global marketplace.
- Economy is information, ICT and biotechnology based.
- Rise of individualism linked to emancipation of women in the labour market, new attitudes to contraception and abortion, and greater financial independence.

Examiner tip

You can find evolutionary population pyramids and a wealth of demographic data for almost all countries at **www. nationmaster.com**

- Greater environmental concerns for the impact of increased numbers on resources.
- Rise of non-traditional lifestyle, such as same-sex relationships.
- Rise of childlessness and delayed pregnancy.

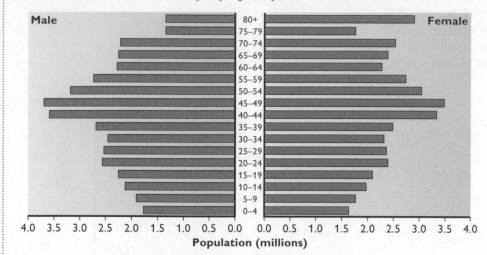

Figure 8 Germany: population pyramid for 2010

Knowledge check 7

Give three outstanding characteristics of the German population pyramid in Figure 8.

Implications of population change

World population is growing **exponentially** or **geometrically** — the number increases at a constant proportion of the number at a previous time. It appears as an increasingly steep curve on normal graph paper and a straight line on logarithmic graph paper. Table 2 tabulates our concerns by continent, using forecast data for 2050.

Knowledge check 8

Can you describe the differences between the rows for Africa, Asia and Europe?

To cope with population growth there have to be resources to support the population. The ability to support a population is measured by the **population–resource ratio**, an out of date attempt to link the ability of a population to support itself from its own resources and by trading.

Table 2 Population growth, 1950–2050

Continent	Population (millions)			Share of world population (%)			Change (millions)		Share of world change (%)	
	1950	2010	2050[1]	1950	2010	2050[1]	1950–2010	2010–2050[1]	1950–2010	2010–2050
Africa	221	1,030	1,803	8.8	15.0	20.3	809	773	18.6	38.2
Asia	1,402	4,157	5,222	55.6	60.4	58.6	2,755	1,065	63.1	52.6
Europe	547	739	633	21.7	10.7	7.0	192	−106	4.4	−5.2
Latin America	167	585	767	6.6	8.5	8.6	418	182	9.5	8.9
North America	172	344	448	6.8	4.9	5.0	172	104	3.9	5.1
Oceania	13	37	45	0.5	0.5	0.5	24	8	0.5	0.4
World	**2,522**	**6,892**	**8,918**	**100**	**100**	**100**	**4,370**	**2,026**	**100**	**100**

[1] = UN 2004 estimates

Sustainable population, once called **optimum population,** is the population that can be supported so that natural resources are not depleted and output is maximised within the prevailing technological, economic and social conditions. It is a theoretical aim for the world and more so for individual countries. The major concern for the future is **overpopulation**, where resources are unable to sustain a population at the existing living standard without a reduction in that population or an increase in the resources available to support it. It is a concern for LEDCs where Malthusian checks such as famine, AIDS and war have an impact.

Underpopulation occurs when existing resources could sustain a larger population without lowering living standards, or where the population is too small to develop the available resources.

It is rarely a concern today, although it is becoming one in some MEDCs where populations are declining naturally. International migration is often encouraged in these circumstances.

Examiner tip

Make sure that you are able to scan tables so that you can quickly extract relevant information. What do the other rows and columns tell you?

Figure 9 Three models of the impact of population growth

Examiner tip

You should be able to draw a diagram of each of the models.

The debate over the ability of the world to support its population has been around since **Thomas Malthus** wrote his theory in 1798. In the twentieth century, **Esther Boserup** (1965) and the Club of Rome (1972) respectively challenged and supported Malthus (Figure 9). **Neo-Malthusians** have suggested that famines in the Sahel and Somalia in 2011 have shown that population is outstripping food supply. The Food and Agriculture Organization (FAO) suggests that up to 800 million people are malnourished. Population growth is accelerating in Africa and impeding development although wars and the AIDS pandemic have slowed the increases. Water shortages, such as in Somalia in 2011, partly caused by climate change, are predicted to become a major issue by 2050. Today, we are more aware than ever of the finite nature of the world's resources. This has increased the importance of **sustainable development** (introduced at the 1972 Global Environmental Summit in Stockholm), and the critical links between population, the economy and the environment. The 2002 World Summit in Johannesburg focused upon sustainable management of the global resource base, poverty eradication and improved healthcare.

1.3 What is the role of migration in population change?

Motives — why do people migrate?

Table 3 provides a basic model of migration at any scale. The balance between immigration and emigration is called **net migration**. It will vary over time, within a country, and between countries. Migration is generally long term and mostly for economic reasons. However, much publicity has been given to refugees and asylum seekers who are often, mistakenly, grouped with economic migrants.

Table 3 Factors that influence individual migration, mobility and access to work

What the job-seeking migrant is looking for — the supply of jobs			What the demand is in an area	
Personal characteristics	Socio-economic factors	Transport	Employers' needs	Demand in the area
Education needed Qualifications for job Work experience of seeker CV/interview skills needed Personality — do I fit?	Work history good/poor Ability to afford child care Age of seeker Gender of seeker Race of seeker Health of seeker	Private transport Public transport Accessibility of job Cost of transport Housing costs	Discrimination, which is real or perceived, based on: • Work experience of applicants • Post code • How long unemployed/ employed • Age/gender/race in area of recruitment	How many jobs in the area? Is it an area of employment growth? Occupation/skills Whether full or part time Quality of workers in area

Table 3 shows that there are a whole host of other factors that will come into play when a person is looking to migrate for economic reasons. In MEDCs these factors will explain why some people find it easier to migrate for work than others.

Voluntary or forced migration?

Most migration is the result of free choices made by the migrant, the head of a migrant household, or a group. However, according to the UNHCR, 22 million people have been forced to leave their homes and/or country. See Table 4 for examples.

Table 4 A typology of international migration

Voluntary	Examples	Forced	Examples
Between MEDCs	Italians to Bedford in the 1950s: brickworks. Poles to work in UK after 2004.	Between MEDCs	Germans from Hungary to West Germany in 1945. French from Alsace to France after Franco– Prussian war in 1871

Voluntary	Examples	Forced	Examples
Skilled labour	Financial experts to New York and Singapore	n/a	n/a
Between LEDCs	Lesotho to South Africa	Between LEDCs	Hutus from Rwanda to DR Congo. Palestinians to West Bank and Gaza
LEDC to MEDC	Bangladeshis to UAE	Sex workers and child trafficking	Nigerians to UK
Labour migrants	Caribbean to UK in 1950s. Mexicans to USA	Labour migrants	Sex industry workers from Eastern Europe
Refugees	Montserrat to Antigua following volcanic eruption	Refugees	Ethnic cleansing in Southern Sudan in 2010. Libyans to Italy in 2011
Asylum seekers	Afghans to Australia/UK, Iraqis to UK	Asylum seekers	Kurds from Iraq to Italy

Models of migration

Ravenstein's laws

Ravenstein's laws are based on Britain in the 1880s. He gave migration the following characteristics:
- it is short distance
- it is step-by-step
- longer distances are travelled to major centres
- there are reverse flows
- rural people are more inclined to migrate than urban people
- females move more than males
- most migrants are adults
- large towns grow more by migration than by natural increase
- volume grows as industrialisation progresses
- the major flow is from agriculture to industrial towns
- the major cause is economic

Lee's model

According to Lee, people:
- assess and perceive the destination
- assess conditions where they are
- look at obstacles between the two, such as distance and cost
- consider personal circumstances

Stouffer's intervening opportunities model

This model (Figure 10) attempts to explain why migrants settle for locations other than their original, intended destination.

> **Knowledge check 9**
>
> Give alternative examples to those in Table 4.

> **Knowledge check 10**
>
> Do these 'laws' apply today? If so where and why? If they do not, why is this so?

Figure 10 Stouffer's intervening opportunities model

Todaro's model

Todaro's model is based on two principles:

- Economic factors are the most important of the push–pull factors.
- Migration is most likely where urban incomes are greater than rural incomes.

The decision-making model

Knowledge check 11

Can you find your own examples to support the points made in Figure 11?

Figure 11 The decision-making model

Migration at a national scale: the UK

In the nineteenth century people migrated to the coalfields, for example to South Wales from rural Wales and Ireland. In the first half of the twentieth century there was movement towards London and the South East. Since the 1930s, the process of **suburbanisation** has seen movement to the outer edges of cities. **Counter-urbanisation** in the late twentieth century continues today, and is a net movement of people away from cities to small towns and villages. There are other forms of migration: retirement migration to southwest England, emigration to live in the Dordogne or Spanish Costas following second-home purchase, and student migrations to university towns and then away again on graduation. There is significant **re-urbanisation** (migration back into cities) by mainly high-flying single people and the wealthy.

In 2011, 11.5% of the UK population was born outside the country: 684,000 from India, 521,000 from Poland, 419,000 from Pakistan represent the top three countries of birth. Ireland, Germany, South Africa, Bangladesh, USA, Jamaica, Nigeria, Kenya and Zimbabwe have all sent over 125,000 persons to the UK. The issues in the UK are:

- Media portrayal of immigration using emotive language, for example, 'flood', 'tidal flow' and 'swamping' of services.
- Difficulty of distinguishing economic migrants from refugees and asylum seekers.
- Migrant communities encouraging chain migration of relatives to the same area.
- Issues of controlling the UK border.
- Pressure on housing, education, health service and social services.
- Much of the debate is clouded by the issues of refugees and asylum seekers.

> **Knowledge check 12**
>
> Are the reasons for migration identical for each country of origin? What factors might explain these patterns of migration?

The impact of international migration

More countries are being affected by international migration as the numbers migrating grow. There is more variety among migrants and no type dominates the flows as in the past. More women are labour migrants in their own right. These details pose new challenges both for receiving and exporting states. More migration is short term as states place limits on work permits. Long term migration has declined because of political pressures from the host population, the lack of low-skilled jobs, and tightened entry requirements such as the UK points and quota systems — the **Highly Skilled Migrant Programme**. Although people migrate for work, many return home later.

Impact on the country accepting the migrants

- Demographic replacement for low natural change and declining population, for example Turks to West Germany in the 1970s, Croatians to Switzerland and Iranians to Sweden (although some sources may class these as refugees). In 2002, Germany recognised the need for 50,000 migrants per year as a demographic replacement for retiring Germans.
- Labour force needs are satisfied, e.g. Caribbeans to the UK in the 1950s, Portuguese (Madeirans) to Guernsey in the 1990s and more recently Latvians. Some of these movements become permanent because of marriage but many comprise **guest workers** — people recruited for a period of time such as South Asians to the UAE, or employees on assignment such as UK workers from Barclays Bank and Microsoft working in New York and Seattle.
- Increased pressure of population on resources, such as food and land, housing and social resources. This is the argument made by right-wing groups in the UK.

- It helps to create a **multicultural society**, e.g. Toronto. The religious mix, retailing, restaurants and music of the host country may alter.

Knowledge check 13

Are there areas dominated by an immigrant group in a city that you know? Why does this happen?

- Areas may become dominated by an immigrant group, e.g. Turks in Cologne, Caribbeans in New York and New Jersey, and the Chinese in the west coast cities of the USA creating vibrant Chinatowns. There are regional variations, as illustrated in the UK: in Yorkshire & Humberside and the West Midlands the dominant groups are Pakistanis and Indians; in Wales and Scotland they are Poles and Indians; in the northeast, Indians and Germans; whereas in the northwest they are Pakistanis and Poles. In contrast, the southeast is dominated by Indians and South Africans and the southwest by Poles and Germans. Polish-born residents are either the biggest or second biggest foreign-born populations in eight regions of the UK, London has 123,000 Polish-born residents, 24 per cent of the UK total. The fifth biggest foreign-born population in Wales, comprising 6,000 people, is Filipino. Brent has 54% foreign-born people, the largest proportion in the UK.

Examiner tip

Be able to illustrate regional variations in foreign-born populations in the UK using figures.

- There are issues of race relations and **segregation** such as the divide in Blackburn between immigrant peoples on one side of town and the long-standing residents on the other. **White flight** and **ghettoisation** may be further outcomes.
- Political reactions such as BNP, National Front in France, Austrian Freedom Party, Vlaams Blok in Belgium.
- Gender concentrations, e.g. south Asian men in the UAE and Mexican women in the USA.

Knowledge check 14

What do you understand by the term 'segregation'?

- Illegal labour on low wages, e.g. Mexicans on Californian farms, and some of the gangmaster cases in the UK.

Impact on the country sending the migrants

- Slowing down of natural increase because the fertile migrate.
- Old-age society, as young adults leave.
- Slightly decreased pressure on resources, e.g. after the Irish potato famine.
- Fewer people to engage in agriculture and produce food.
- Receipt of **remittances** from workers to family back home. However, the flow of remittances can decline, e.g. from USA to Mexico in 2009, resulting in hardship for families back home. Remittances fell by 18% in 2009.
- **Return migration** with new skills and prospects, e.g. taxi drivers and chefs to the Mediterranean resorts.
- **Westernisation** and **cultural imperialism** reinforced by returnees.
- Solution to a political or racial issue, e.g. ethnic cleansing in the Balkans, and Ugandan Asians in the 1970s who left for the UK.
- Loss of skills, e.g. IT experts moving to USA from India.
- Pensions outflow where retirement migration is taking place, e.g. from the UK to Spain and Dordogne.
- Return migration during hard times, whether economic or political, in the receiving countries.

Return migration

Economic downturns such as the 2008–09 credit crunch have the effect of encouraging many economic migrants to return to their country of origin. In 2009 it was noted that professional Nigerians and Ghanaians working in the UK were returning home, a phenomenon called **brain gain**. Similarly, there is a return flow from the USA to Mexico resulting in a reduction of income for Mexican families. For other labour migrants, returning home was part of their original plans once they had made money. Throughout the past 40 years there has been a steady return migration from Germany and Switzerland to the countries of southern Europe. Many returnees had acquired the capital to set themselves up in business in their home area.

Examiner tip

Always make it very plain which type of migration you are discussing and at what scale — international, inter-regional or local.

Migrant labour: economic migration

In 2001, Jersey had 38,662 persons born outside the island. British born formed 36% of the population and were declining in numbers. Portuguese and especially Madeiran born were 6.4% of the total population; Irish born were 2.6%. The population for the Portuguese was typical of labour migrants, with 54.6% aged between 20 and 39 years old. Males slightly outnumbered females in these age groups. Among the Irish, females outnumbered males, especially in the 20–29 age group.

These differences probably reflect a number of factors:
- The Irish have a language affinity with the islanders and were earlier migrants.
- Distance from the home country has meant that males were more likely to be the dominant Portuguese migrants. In addition, many Portuguese migrants come from islands such as Madeira and the Azores that are even further away.
- It is mainly a singles population judging from the small number of dependents and retired migrants.
- Both Ireland and Portugal have higher unemployment and lower living standards than Jersey.

The Portuguese and Irish are short-term migrants, often returning home at the end of the tourist season. Both groups are employed in lower-skilled work, particularly in hotels, although some of those who have settled or keep returning to the island have risen to the status of managers in the holiday industry. UK immigrants are affluent people seeking a tax haven and skilled workers attracted to work in the finance sector and government employment, such as education, and are less likely to be recent migrants. Among the consequences of the Portuguese presence are that:
- shops in St Helier have Portuguese speaking assistants to serve the immigrants
- the BBC in Jersey has Portuguese language broadcasts
- the Catholic Church has services in Portuguese

Guest workers make an important contribution to the national economy in Switzerland (26.5% of the workforce in 2007), Germany (7.5% of the workforce) and the Gulf States such as UAE, where many males have migrated from South Asia.

Table 5 summarises some of the costs and benefits of migrant labour to both the source and destination countries.

Knowledge check 15

Do you have a local example of an activity being dominated by economic migration?

Table 5 Issues resulting from labour migration

Source country		Destination country	
Economic costs	**Economic benefits**	**Economic costs**	**Economic benefits**
Loss of young adult labour to source country	Less underemployment	Costs of educating children	Many less desirable jobs taken in host country (fruit picking)

Source country		Destination country	
Economic costs	Economic benefits	Economic costs	Economic benefits
Loss of skilled labour may slow development	Return migrants (reverse or counter-migration) bring skills back home	Some skills may be bought cheaper, so home people are disadvantaged	The host country may gain skilled labour at little cost; can fill skills gap (e.g. programmers from India to USA)
Regions of strongest out-movement suffer from spiral of decline (e.g. parts of inland southern Italy)	Funds repatriated to source country	Funds repatriated to source country; pension outflows following retirement to home country	Some retirement costs transferred back to source country
Loss of skilled labour may deter inward investment	Less pressure on resources such as food	Slightly greater pressure on resources	Dependence of some industries on migrants (e.g. food processing in West Sussex)
Encourages more to migrate, with effect on social structure (e.g. Ireland in the 1840s)	Reduced population density; lowers birth rate because young tend to migrate	Discrimination against ethnic minorities, political unrest and opposition from parties of the right (e.g. in Austria)	Creation of multi-ethnic, multicultural society – building cultural facilities
Dominance of females left behind	Funds sent home can provide for improved education and welfare	Dominance of males, especially in states where status of women is low (e.g. Gulf States)	Greater awareness of other cultures (e.g. Caribbean music, literature and religion in the UK)
Non-return can unbalance population pyramid	Retirees building homes	Loss of aspects of cultural identity, especially among second generation	Providers of local services (e.g. Turkish baths and UK newsagents)
Returning on retirement is a potential cost	Some able to help develop new activities (e.g. tourism)	Creation of ghettos (e.g. Turks in Berlin); schools dominated by immigrants	Growth of ethnic retailing and restaurants (e.g. Rusholme, Manchester)

1.4 Refugees and asylum seekers

You need to distinguish all of the following:

- **Asylum seekers** —the formal application by a refugee to reside in a country when they arrive in that country.
- **Economic migrants** — people seeking opportunities and legitimately entering a country.
- **Illegal immigrants** — people who enter a country without authority and hope to remain.
- **Internally displaced persons** — persons or groups who have been forced or obliged to flee or to leave their homes or places of habitual residence, as a result of, or in order to avoid the effects of, armed conflict, situations of generalised violence, violations of human rights or natural or human-made disasters, and who have not crossed an internationally recognised state border.
- **Refugees** — those genuinely fleeing from persecution and harsh political regimes. A refugee is defined by the United Nations (UN) as 'a person unable or unwilling to return to his/her homeland for fear of persecution based on reasons of race, religion, ethnicity, membership of a particular social group, or political opinion'.

- **Stateless persons** — any people who are not considered as nationals by any state through its nationality legislation or constitution. There were 12 million in 2010.

Cases such as Afghans today trying to board lorries passing through the Channel Tunnel illustrate how difficult making this distinction can be. Are they refugees seeking asylum or are they illegal immigrants?

The UN projection is that the population of Europe will decline by 5% by 2025 and therefore the labour force will have to be replaced either by a higher rate of natural increase or by the immigration of economic migrants, refugees and asylum seekers. There have been refugees throughout history. Examples include:

- Persons displaced in post-1945 Europe — for example, Germans in Moldova.
- Movements between India and Pakistan in 1947 upon independence of the two countries.
- Vietnam in 1975 when 800,000 boat people fled — some are still in Hong Kong.
- The 2011 flight from drought and war in Somalia to refugee camps in Kenya such as Dadaab.
- Afghans to Pakistan — the most costly relative to GDP — 710 refugees for every $ of GDP. 30% of the world's refugees are from Afghanistan and 96% of those are in Pakistan.

Examiner tip

Always define what you mean when you use terms. Do so in the introduction to essays.

Refugees

The UNHCR estimated that there were 43.7 million people displaced forcibly from either their home areas (**internally displaced persons — IDPs**) or their home countries. In 2010 there were 15.4 million refugees. Most refugees go to their neighbouring countries and 80% are in developing countries. Europe only hosts 15% of the world's refugees. UNHCR's African refugee emergency areas are concentrated in sub-Saharan Africa where there were 2.2 million in 2010.

Norway, Sweden, Finland and Switzerland have a long tradition of accepting refugees and asylum seekers who are unable to stay in their intended country of refuge. Canada adds gender persecution to its criteria for accepting refugees. Figure 12 shows the major source countries of refugees, asylum seekers and internally displaced persons in 2010.

Figure 12 Major source countries of refugees, asylum seekers and internally displaced persons, end 2010

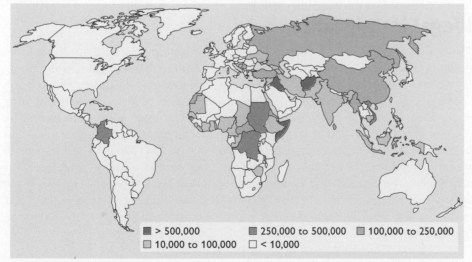

> 500,000   250,000 to 500,000   100,000 to 250,000
10,000 to 100,000   < 10,000

Rwanda and Burundi

In 1994 in Rwanda, the **genocide** of at least 500,000 people displaced forcibly as a consequence of civil war between the Tutsi and Hutu peoples resulted in the movement of 1.75 million refugee Hutus to neighbouring countries. As Hutus left, a counter flow of 700,000 Tutsi refugees began. The UNHCR encouraged repatriation and about 160,000 Hutus returned. But the high birth rate among the refugees actually meant that the population remaining outside Rwanda stayed the same. The refugees were housed in vast camps near the Rwandan border in DR Congo and Tanzania. In 2003 the refugees in Goma were forced to move again by the eruption of Mount Nyiragongo.

In Burundi, a similar tribal conflict commenced in 1993 and was made worse by the flight of the Hutus from Rwanda. Many were forced back or fled to Tanzania. The political situation in eastern Congo, where civil war was raging, resulted in the Congolese forcing 650,000 to return to Rwanda and Burundi. By 1996, 1.7 million Rwandans and 200,000 Burundians were in camps. At the end of the crisis in 1997, 600,000 were repatriated from Tanzania and the Congo but 215,000 remained unaccounted for.

The crisis resulted in the deaths of over a million people and in environmental degradation around the refugee camps, which were denuded of timber. War and flight had halted all development. Even in 2009 many still lived in poverty in camps. In 2007 300,000 refugees and 1.7 million IDPs decided to return home, mainly to Burundi and DR Congo.

Refugees in the UK

In 2010, only 20% of the estimated 15.2 million refugees reached developed countries. In the same year the UK hosted 238,150 refugees whereas Germany hosted 594,269. The UK is not the most receptive country and is only tenth in Europe for the number of refugees as a percentage of its population. Therefore, you must check the accuracy of media statements on refugees. The majority (some estimate 1 in 20) live in London. Newham, Enfield, Haringey, Redbridge and Waltham Forest are the boroughs with the highest totals (including asylum seekers).

Examiner tip

The digital magazine *The New Londoners* (**www. thenewlondoners. co.uk**) is a good, if biased, expression of the feelings of migrant communities in London.

Illegal immigrants

There are an estimated 2.6 million illegal immigrants in Western Europe, and this figure is exacerbated by the flow into southern Europe following the 'Arab Spring' revolutions in 2011. Nearly all countries are attempting to tighten immigration controls (e.g. UK Border Agency), often in the wake of xenophobic (right-wing nationalist) attacks on refugees and 'illegals'. In 2009 there were estimated to be 725,000 illegal immigrants in the UK although the very nature of illegal immigration means that the figure could range between 524,000 and 947,000. The destinations of illegal immigrants are not easy to ascertain. They spread widely and are probably distributed where the illegal employment opportunities are best — in low-grade agricultural work such as vegetable picking, the hotel industry and labouring jobs. London is disproportionately affected. In 2007, estimates showed that 518,000 illegal immigrants lived in London. 188,00 illegal immigrants have been deported since 1998. Deportations in 2010 cost £150,000 per deportation.

Asylum

The number of asylum seekers is invariably larger than the number who request refugee status and has increased steadily since 1970 when many countries decided to curtail immigration. One catalyst was the German realisation that Turkish guest workers were unlikely to return home. The collapse of Yugoslavia in 1992 and the subsequent civil unrest in the region resulted in a rise in asylum seekers from Bosnia-Herzegovina, Croatia, Serbia and Montenegro. Figure 13 shows the origins of global asylum seekers in 2010, which totalled over half a million worldwide.

There were 17,790 applicants for asylum in the UK in 2010, a 13% decline in a year. Of these, 17% were granted asylum, 8% granted either humanitarian protection or discretionary leave and 75% were refused. In the same year, 9,850 persons were removed or departed voluntarily from the UK. Table 6 places the UK in its European context.

Figure 13 Countries of origin of new asylum seekers in 2010

Table 6 Asylum applications in Europe, 2010

Country (world ranking for numbers applying)	Number	Applicants per 10,000 inhabitants	Percentage granted	Percentage of European (EU) total
France (2)	47,790	8	23	16.3
Italy (14)	10,096	2	31	11.7
Germany (3)	41,350	5	22	10.5
UK (6)	22,190	4	24	9.7
Sweden (4)	31,820	34	23	15.2
Greece	10,270	9	36	7.7
Switzerland (not EU) (8)	13,500	nd	nd	
Belgium (7)	19,940	19	17	6.2
Netherlands (9)	13,350	8	46	5.9
Norway (not EU) (12)	14,430	nd	nd	
EU 27	235,930	5	23	100

Knowledge check 16

Define refugee and asylum seeker.

Examiner tip

The different results of the figures for asylum applications in the UK in 2010, shown in Figure 13 and Table 6, are taken from contrasting sources. Be aware of the possibility of contradictory information in such statistics.

The flow of asylum seekers is to 44 industrialised countries. The pattern can be explained as follows:

- Proximity — Italy for Albanian and north African applicants, Greece for Balkan, Turkish and Kurdish applicants.
- Receptiveness — Scandinavian reputation for compassion and receiving migrants.
- Neutrality — Switzerland and Sweden.
- Perceived job opportunities — Germany, France, UK.
- Links as a result of colonialism — Zimbabweans to UK, West Africans to France.

Reasons for increased prominence of asylum seeking

- Pressures to migrate from the poorest states are increasing due to a combination of economic decline and political instability (e.g. Zimbabwe, and Libya in 2011).
- Improved media enable people to learn more about destinations and the means to reach them.
- Communities already exist in destinations and may encourage extended family members to join them. This is **chain migration**.
- The variety of transport modes encourages people to try.
- Gangs of human traffickers prey on would-be migrants and offer passage to a new life.
- Destination countries find it difficult to distinguish between those fleeing from threats to their life and liberty, and those trying to escape poverty and improve their quality of life.

Examiner tip

Beware of bias, especially from the media.
Appreciate the position of newspapers when it comes to reporting this issue. Collect examples of differing views from the media.

Costs of asylum seeking in destination countries

Costs associated with asylum seeking include:

- housing — 18,000 received housing support in 2010
- social services
- schooling — in 2010 21% were under 17 years old.
- welfare payments (note that they are not eligible for main welfare benefits in the UK)
- 'improved' border controls at points of entry
- dispersal costs to areas where housing exists — Liverpool, Birmingham, Manchester and Glasgow are the main dispersal cities
- policing of demonstrations against these people by right-wing extremists

These costs have to be paid for from taxation although it is hoped that those granted asylum will become taxpayers. Asylum seekers often wish to live in the heart of a country such as the capital city and do not want to be dispersed while their application is processed. Total costs for refused applicants, which can rise to £50,000 for a family, are far higher than for those families granted asylum (up to £12,000).

Policy reactions

(1) Limiting seekers at source with more difficult visa requirements.

(2) Pre-boarding arrangements to prevent departure unless the passenger has a return ticket.

(3) Prevention of illegal crossings by physical constraints, e.g. Channel Tunnel.

(4) Fast-track procedures for the genuine cases.

(5) Accepting those with skills, often using a points system.

(6) Maintain third country policies as used by Germany, whereby people entering the EU in Bulgaria, the Czech Republic and Poland, and intending to go to Germany are subject to German regulation.

(7) Temporary protection followed by repatriation once crisis subsides — used for Bosnians in 1996–97.

(8) Policies of NGOs aiding governments in the countries of both origin and destination to address the deep-seated causes of poverty.

Stateless people and ethnic cleansing

The UN Universal Declaration on Human Rights states that everyone has the right to a nationality, yet UNHCR estimates that there are 12 million stateless people worldwide who are denied the right to have a home and to work. The largest European group is the Roma Gypsy population: there are 12 million in Europe (Romania 1.8 million and Bulgaria 700,000). In some cities they live in ghettos and their life expectancy, on average, is 15 years less than that in the host nation. In 2004 1 million became EU citizens and technically no longer stateless.

The break-up of Yugoslavia resulted in some groups being declared stateless for political and nationalistic reasons (e.g. Serbs in Croatia and Macedonia). The result has been **ethnic cleansing** — the forced and voluntary movement of peoples to states that are culturally compatible (having the same religion and/or language).

Other cases of statelessness include 3 million Palestinians in Gaza and the West Bank, and Myanmari Muslims.

1.5 Changing gender structures

Gender is a generic term that refers to both males and females. Women's role in society has demographic, economic, environmental and social causes and effects, including attitudes to work and child rearing, attitudes to female children, and attitudes to migration. There have been imbalances in gender roles because of historic causes. Traditionally in northeast England there has been less work for women because the main employment was in coal mining and shipbuilding. In contrast, Lancashire had a tradition of female labour in the textile industry. These differences are less marked in the twenty-first century.

Gender and the demographic transition

In countries at Stage 2 women are regarded as mothers for children due to the high infant and childhood mortality rates. In much of sub-Saharan Africa women are the cooks and providers of agricultural labour. In Guinea Bissau women still collect 90% of the water. Depending on the country, sub-Saharan African women provide 60–80% of the agricultural labour. Others work in the informal, unregulated economy where wages are low. Women are more likely to be illiterate because they have not

---

Examiner tip

This topic is one where you must be able to keep up to date because good case studies appear in the quality media all of the time. Amnesty International is a good source of new information. Remember that new information that you have gathered gains credit.

Knowledge check 17

Define a stateless person

Examiner tip

In any question related to gender you may refer to males as well as females.

had the opportunity to be educated. There are **matrilineal** societies where women are more dominant and property passes from mother to daughter, for example the Minangkabau people in Sumatra, Indonesia, a country at Stage 3.

Many Stage 3 countries use women as cheap labour. In Asia up to 50% of women provide agricultural labour and in some newly industrialised countries (NICs), such as Thailand, they are providing labour in food processing and other industries. The demographic transition has indirectly led to more women being educated.

In LEDCs women may lack schooling and the full and legal rights of citizenship. Some depend on children for status and security. Policies to improve women's rights include:

- Giving girls a good education, which attempts to empower women outside the home (Figure 14). Many girls, particularly in Africa, do not have equal access to schooling and so there is a particular need to provide secondary education in those countries.
- Providing employment or small business opportunities.
- Reducing the impact of patriarchal societies where children are regarded as a statement of virility.
- Reproductive health clinics to assist women to value their lives.
- Emphasis on contraception for both men and women.
- Getting tribal leaders to promote lower birth rates.
- Normalising the biological sex ratio, which at birth is 102–106 girls for every 100 boys. In China it is 120:100, Armenia 115 and India 108.

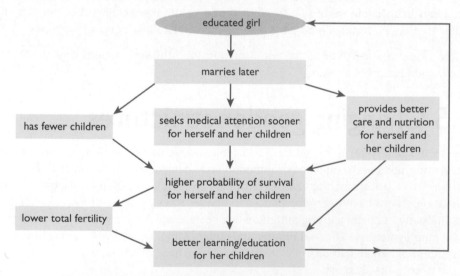

Figure 14 Possible demographic impacts of educating girls

UNI Women, the United Nations Entity for Gender Equality and the Empowerment of Women, founded in 2010, developed out of UNIFEM, the United Nations Development Fund for Women, founded in 1976. It is fostering empowerment through gender equality and has focused on the role of women in the spread of AIDS in India. Currently, among 15–24-year-olds who have the AIDS virus worldwide, 53% are women. Counteracting the AIDS pandemic is a gender issue because it affects both men and women.

> **Examiner tip**
>
> When discussing gender issues it is advisable to know more than just the Chinese single child policy and Indian female infanticide.

The Millennium Development Goals' Target 1B addresses women's issues where it aims to have full and productive employment and decent work for all. In addressing the issue of hunger and the 50% of people living on under US$1 a day it addresses the needs of women. In 2011 'UN Women' admitted that Target 1B was unlikely to be met by 2015.

Stage 3 still reflects a world dominated by men. Industrialisation and industrial urbanisation results in employment dominated by men. Transport developments such as railways and trams provide opportunities for men to travel to work from the suburbs. Nevertheless, the textile industry did provide opportunities for women. The World Wars were a catalyst for changing attitudes to women at work. Suddenly, women were needed to keep factories operating (not least munitions factories), to work in the service sector (e.g. bus conductresses) and in agriculture (the Land Army in the Second World War).

The USA is a good example of a country at Stage 4 of the demographic transition where, in 2010, 72 million working women made up 49.9% of the labour force. They worked in the following areas:
- 40% in management and professional jobs
- 32% in sales and office work
- 21% in other service sector jobs
- 5% in industry and transport
- 1% in resource related jobs

Female employment did show racial variations: whites and Asians held 41% and 47% respectively of female management and professional positions whereas Afro-Caribbeans and Hispanics held 32% and 33% of sales and office jobs. The types of occupation still reflect a degree of gender bias. The 10 dominant occupations were: retailing (3.5 million), nursing, including psychiatric and home help (2.6 m); secretaries and administrative assistants (3 m), school teachers (2.3 m); cashiers (2.2 m); waitresses (1.4 m); maids and housekeepers (1.2 m); customer services (1.2 m); child care (1.2 m).

Knowledge check 18

In Stage 4 countries, in what sectors of the economy do most women work?

Women's enhanced role is partly due to the available educational opportunities. One third of both genders have degree level qualifications.

Stage 5 is difficult to relate to gender development because those countries with declining populations, such as Italy, are not ranked very highly on the HDI (Table 7). Stage 5 countries have a low birth rate, and a high degree of female emancipation and contraceptive use, which when placed alongside delayed pregnancy, small family size and same-gender unions, is a distinctive aspect of the stage. In Stages 4 and 5 women live longer and outnumber men in the older age groups. War losses affect the balance between males and females, for example in Germany after the two World Wars. **Glass ceilings** that prevent women, even in the most gender-equal societies, from attaining equality of opportunity for the top jobs still exist. Nevertheless, female employment is rising in Europe (Figure 15).

The UNDP measures gender inequality using the Gender Inequality Index (GII). The GII is based on indicators of: women in the workforce, educational attainment of women, women in parliament, adolescent fertility and maternal mortality, where 0.0 is equality and 0.99 complete inequality. The classification has been related to the

2010 Human Development Index on Table 7. HDI uses the dimensions of: (i) a long and healthy life indicated by life expectancy of men and women; (ii) knowledge indicated by adult literacy rates; and (iii) a decent standard of living indicated by estimated earned income.

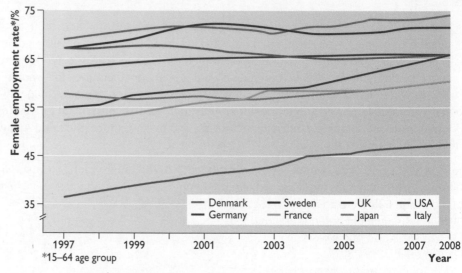

Figure 15 Changes in female employment in Europe

Gender and migration

Migration affects gender balance in the areas of origin and destination. For every 2.5 Mexican males entering the USA, only 1 woman makes the same journey. What is the effect on the areas of origin?

Table 7 Selected country data on Gender Inequality Index (GII)

HDI rank 2010	GII rank (and value) 2011	Life expectancy at birth in years, 2005–2010		Adult (15 and over) literacy rate (%), 2009–2011		Maternal mortality rate per 100,000 live births 2008
Very high human development		F	M	F	M	
1 Norway	6 (0.075)	82.5	77.8	100	100	8
4 USA	47 (0.299)	80.4	75.2	100	100	24
7 Netherlands	2 (0.052)	81.9	77.5	100	100	9
13 Switzerland	4 (0.067)	84.2	79.0	100	100	10
23 Italy	15 (0.124)	83.5	77.5	99	99	5
26 UK	34 (0.209)	81.6	77.2	100	100	12

Knowledge check 19

Use Table 7 to comment on the variations of male/female opportunity.

High human development						
51 Cuba	58 (0.337)	76.2	80.4	100	100	53
56 Saudi Arabia	135 (0.646)	75.3	70.9	81	90	24
68 Kazakhstan	56 (0.334)	72.4	61.6	100	100	45
73 Venezuela	78 (0.447)	76.8	70.9	95	95	68
Medium human development						
101 China	35 (0.209)	74.8	71.3	91	97	38
123 South Africa	94 (0.490)	49.7	48.8	87	91	410
134 India	129 (0.617)	66.4	63.2	51	75	230
Low human development						
125 Pakistan	112 (0.721)	65.8	65.2	40	69	260
154 Yemen	146 (0.769) (lowest rank)	64.3	61.1	72	96	210
161 Uganda	116 (0.662)	52.2	50.2	62	81	430
173 Zimbabwe	118 (0.583)	42.6	44.1	89	95	790
175 Mali	143 (0.712)	56.1	52.1	18	35	830
180 Sierra Leone	137 (0.756)	44.1	41.0	30	53	970
185 Burundi	89 (0.478)	51.0	48.1	61	73	970
187 DR Congo (lowest rank)	142 (0.710)	47.7	45.2	57	77	670

(after UNDP Human Development Reports 2007–2010)

Gender concentrations

Park and Burgess (the same person who identified the model) noted in the 1920s **moral regions**, the areas of prostitution and vice in Chicago. Gay 'villages' can develop — for example, in San Francisco, Castro is the district where gay men live. Male prostitution is found on Polk Street. Female prostitution is found at Mission and 18th Street. Valencia Street is the core of the lesbian community.

Canal Street in Manchester is associated with the gay community, but as a recreational area rather than where they live.

There are areas of cities that can contain more of one gender. The nature of the housing market and the type of housing enables single females to purchase properties in an area. Central Southsea is a ward with a disproportionate number of households occupied by 20–39-year-old single females. University towns often contain areas where females outnumber males: females form 57% of students in the USA. Working women are in the majority in Gwynedd, Pembrokeshire and Anglesey whereas they are a significant minority in Tower Hamlets (40%) and Southwark (43%).

Examiner tip

Try to be able to quote the figures for one country from each of the four gender columns on Table 7 as illustrations of points concerning gender variations.

Knowledge check 20

Can moral regions be found in UK cities?

Examiner tip

Find other examples of gay villages and where they are located within an urban area. Why are they there?

Knowledge check 21

Can you explain why women dominate employment in some occupations and not others?

Examiner tip

The key aspects of gender are that the roles of women change as countries develop, and that demographic transition and migration affect gender balance, both in the countries of origin and destination.

1.6 What are the demographic challenges facing countries?

Impacts and issues in MEDCs

In 1901 in the UK, life expectancy for men and women was 45 and 49 respectively. In 2001, it was 75 for men and 80 for women. Women aged 65 in 2009 could expect to live a further 20.2 years, whereas men of the same age could expect to live a further 17.6 years. In 1998, the over 60s outnumbered the under 16s for the first time. The proportion of the elderly in the UK is rising faster than anywhere else in Europe.

1 Ageing society

In 2010, 17% of the UK population were over 65; those over 85 were 2.3% of the population and by 2035 the figure for the over 65s is forecast to be 23%. When you retire you will be part of the 29% over 65. The elderly are dominated by females with a ratio of 2 females to every man at age 89. The baby boom children of the post-1945 period are now retired. The elderly often migrate on or just before retirement — **retirement migration**. This may be by **downsizing**, moving to a smaller property, or driven by the desire to live in one of the retirement regions of the UK such as the southwest. Overseas retirement migration has grown although it may be slowed by the 2008–09 credit crunch's effects on private pensions. Some concentrations of older persons remain in the inner city where they have been trapped by low incomes and the dependence on social services and pension credits. Out-migration of the young from areas such as mid-Wales has also led to an apparent concentration of the elderly. Some concentrations are caused by **return migration** as those who left to work in the cities 40 years ago now return to their roots.

The effects of an ageing society on policies can take many forms:
- The conversion of larger properties to nursing and care homes. New purpose-built facilities may also be developed. However, some local authorities are seeking to restrict conversions of properties to care homes, e.g. Bournemouth.
- Companies to cater for the housing needs of the elderly have been established, e.g. McCarthy & Stone.
- Policies to employ the elderly, e.g. B&Q.
- Importance of the grey vote, reflected e.g. in policies on law and order.
- Raising the retirement age — the UK government has already raised the retirement age for women to 65 from 2018, to 66 for all in 2020 and eventually to 68 in 2044 (i.e. for all those born after 1976). In 2010 the minimum retirement age for private pensions rose from 50 to 55.
- Increased leisure facilities, e.g. Saga holidays.
- More businesses geared to the elderly, e.g. RIAS insurance for the over 50s.
- Pressures on health services ranging from care to geriatric beds in hospitals.
- Local economy and retailing changes to cater for an elderly population spending what is called the 'grey pound').
- Growth of new educational provision, e.g. University of the Third Age.

Conversely:
- Fewer people will be moving into work.
- There may be less demand for teachers and schools in areas where the birth rate is low.
- There will be fewer people in work to support the dependent elderly.

Policies to alleviate demographic challenges

Expansionist policies
- Make abortion illegal, e.g. Chile.
- Support for working mothers, e.g. France.
- Improve health care, e.g. UK.
- Restrict availability of birth control, e.g. Ireland and Ghana.
- Give larger family allowances to those with more children, e.g. France today and USSR in the twentieth century.
- Restrict the role of women in society, e.g. Saudi Arabia, and Yemen.

Control policies
- Promotion of birth control, e.g. Singapore.
- Health education in schools, e.g. Netherlands.
- Control family size by controlling number of births per family, e.g. China's one child policy (now lifted in Guandong province).
- Permit abortion, e.g. Russia has 2.7 million per annum.
- Encourage sterilisation with incentives such as TVs and cars for men, e.g. India.

2 Singapore, growth in a confined space

Singapore had a population of over 5 million people in 2010 concentrated in a small area, with no natural resources other than the ability of its 3.7 million resident people. Singapore's policy is to encourage parents to have only one child. To provide the wealth necessary to support the growing population and to improve living standards, the state has invested in high technology and has created a financial centre for the region. It has built a science park and a major conference centre.

Singapore encourages in-migration of skilled labour from Europe and unskilled labour from the surrounding countries to supplement the local workforce and enhance the profitability of the service sector. These activities earn more money to support the building of new towns, e.g. Tampines, to house the population. More recently, the state has sponsored industrial investments overseas in the Indonesian Riau Islands, Malaysia (Johor) and China to gain profits by being the regional headquarters for overseas workbench assembly lines. Low-skilled, low-profit activities have been relocated out of the country so that high-skilled, high-profit activities can take place in Singapore.

The Singapore government has invested heavily in education and has encouraged its educational system to invest overseas. It has developed a global airline (Singapore Airlines) to provide foreign earnings. All of these efforts are underpinned by strong state planning using 5 year plans. As a result, Singapore has developed a resource base to support and sustain moderate population growth.

Knowledge check 22

Define retirement migration.

Examiner tip

Always have figures to quote in support of statements. Just learn the data to the nearest percentage point.

Knowledge check 23

What is a work bench assembly line?

3 India: coping with growth

India, home to 17% of the world's population in 2011, is set to replace China as the world's most populous nation by 2030. India's birth rate has dropped by more than a half from 5.7 children per woman in 1965 to 2.7 in 2010 (or 23/000). Nearly a third of India's people have lowered their fertility to replacement levels. At the same time the death rate has declined from 23/000 to 7/000. In 2001 34% of the population was below the **poverty line** (defined as being able to receive 2,400 calories per day in rural areas and 2,100 in urban areas). Kerala has only 10% poverty whereas Bihar and Rajasthan have over 50%.

Family planning (now called **family welfare**) is encouraged, as is sterilisation (at its peak in 1976 there were 8.3 million sterilisations performed). Sterilisation is seen by some as sexist because it focuses on the female population — 37% of women have opted for sterilisation. Male methods of contraception make up only 6% of contraceptive use. Abortion has been permitted since 1972. In addition, **feticide**, the abortion of girl foetuses, has been common among the middle and upper social groups in richer regions such as Delhi because of dowry issues and in poorer areas such as Rajasthan. Therefore, India has a **gender imbalance** irrespective of population policies. Education, particularly of women (illiteracy was 35% among women and 18% among men in 2011), will assist in lowering the birth rate.

Other policies focused on feeding schoolchildren (43% of children aged 5 are underweight), and building rural homes to raise standards of living. Grain production quadrupled in the late twentieth century thanks to the **green revolution**. Improved health provision halved infant mortality to 30/000 live births in 2011.

India is developing its manufacturing and service sector (call centres and computer software houses) so that it improves its resource base and controls population growth. A third of India's people have lowered their fertility to replacement levels. The task is daunting if the decision to import cereals in 2009 and ban many food exports is a sign of the growing population's demand for food.

Knowledge check 24

What is the gender imbalance in India?

4 Botswana: the impact of AIDS

Botswana is a small, landlocked semi-desert country. In 2010 it had a population of 1.9 million. It has a maternal mortality rate of 190 per 100,000 live births. The infant mortality rate is 48/1,000 and there are 40 doctors for every 100,000 inhabitants. 33% of the population are under 15. Botswana's population would grow more rapidly but for the ravages of HIV/AIDS. Life expectancy is 55 years because of AIDS-related deaths. HIV is prevalent among 18.9% of men and 28.9% of women aged 15–49. Its spread is closely linked to poverty, because of the victims' lack of education, information and access to healthcare. Women are more vulnerable if they are poor and in a society tolerant of extramarital sex, where partners are promiscuous. The consequence is an estimated 170,000 orphans aged under 17.

Examiner tip

Make sure that you know where these countries are in the world.

5 Uganda

In 2010, Uganda's population was 33.8 million of whom 49% were under 15. Growth is slowing and is currently 3.4%. The maternal mortality rate is 5.1/000 live births. The mortality rate for under-5s ranges from 105–194/000 depending on the income of the family. Illiteracy stands at 38% and there are 8 doctors for every 100,000 inhabitants. HIV/AIDS has had a similar effect to that in Botswana although it has declined markedly to 4.3% for men and 6.6% for women aged 15–49. It will take time to change attitudes. Campaigns began against high-risk behaviour, empowering communities through the work of NGOs, educating employers to assist, and providing better access to healthcare. Today, only 8.3% of adults have HIV, because of the success of these measures over the past decade. Therefore, the population still rises rapidly due to improved '**death control**'.

6 Thailand: community-based policy

In 1969, women had, on average, 6.5 children and only 16% used contraceptives. Population growth was 3% per annum. The solution was a nationwide family planning programme which included free contraception, trained family planning specialists, and government campaigns, especially in rural communities. By 2010, 72% of married women were using contraceptives, women averaged 1.8 children and population growth was 0.6% per annum. The population of 68.1 million will continue to rise but the slower rate of increase is the result of a successful community-based, rather than coercive, policy.

> **Knowledge check 25**
>
> What do demographers understand by death control?

Summary

- The basic components of the population system are natural increase and migration which lead to demographic change.
- These changes can be mapped and tabulated.
- Populations change over time should be supported by real data such as that in Table I. Natural population change has been explained by several theories and models of change.
- The consequences of change on people and the environment are not agreed and demographic forecasts range from the gloomy to the more optimistic. There are many different types of migration each with its own causes and consequences for both the exporting and receiving countries or regions.
- The geography of refugees and asylum seekers is very complex and changes rapidly from year to year. People move primarily to improve life chances but for many, the costs outweigh the benefits. It is in this area of study that you must be most aware of the potential for biased treatment by the media.
- Gender may be studied in relation to the demographic transition, migration and the roles of women.
- International variations can lead to marked differences in levels of development (a topic that you might study in G3).

Theme 2: Investigating settlement change in MEDCs

2.1 The features of settlements

The rural–urban continuum

Rural and urban environments are closely linked to each other, for example, through food supply, and work or leisure activities. It is very difficult to distinguish between rural and urban environments using only one feature. In reality there is **a rural–urban continuum** reflecting size, function and density of population.

Size

Many countries employ a **population threshold**, for example 20,000 people, for a settlement to be classed as urban, but the threshold varies between countries and with the character of the area. In sparsely populated areas, such as western Canada or northern Sweden, a settlement of 1000 people may be more urban in character than an agro-town of 30,000 people in southern Italy. The range of minimum populations is from 200 in Norway to 50,000 in Japan. For a UK parish to be classed as rural, the Countryside Commission suggests it should contain fewer than 10,000 people.

Function

Urban areas have a wide range of high-order functions, such as cultural activities, retailing, education, administration and financial services, universities and colleges and specialised retailing. Urban areas, therefore, are **central places** and they have a sphere of influence. In contrast, rural areas usually have comparatively few services, and most of these are concentrated in small market towns. Internet shopping is making such functional distinction less important.

Employment

The main source of employment in urban areas is in the **secondary (manufacturing)**, **tertiary (services)** and **quaternary (research and administration)** sectors. Urban areas attract employees from their catchment areas. There is little, if any, primary activity in urban areas in MEDCs. Rural areas in MEDCs have under 10% working in the primary sector (it can be as low as 2%).

Density

Threshold population densities are a measure of the built-up nature. The usual range is 100–400 people per km². Rural areas in MEDCs are frequently sparsely populated although there can be quite high densities around major cities.

Administration

Many urban areas function as centres of government with larger centres being responsible for several layers of administration.

Character

Urban areas have a social dimension — an urban lifestyle (**urbanism)** that is characterised by pace, stress and a desire to be upwardly mobile. Urban environments are also characterised by a more obvious polarisation, with extreme wealth and acute poverty within small areas. People perceive rural environments as stress free, with a slow pace of life — a rural idyll. However, rural environments in MEDCs contain pockets of poverty.

The **rural–urban continuum** covers a **hierarchy of settlements**:
- **isolated dwellings/farms**
- **hamlet** — a small cluster of dwellings/farms that lack services
- **village** — a settlement in rural surroundings that was once an agricultural settlement. Villages normally contain some services such as a church, an inn and possibly a shop. The term village can be used by estate agents to try and enhance the identity and price of property in former villages engulfed by suburbia
- **large village** — has more services such as antique shops and mini-mart retailing
- **town** — smaller urban area with a range of facilities to serve the area: post office, schools, a range of mainly independent shops
- **city** — a large settlement depending on commerce, manufacturing and service industries. The city region is the area served by a city and usually includes the journey to work area and, importantly these days, the journey to study area
- **conurbation** — urban areas which gradually fused together in the nineteenth century such as Greater Manchester and Leeds–Bradford and the Ruhr area of Germany
- **mega conurbation** — a merging of conurbations whose sheer size is very extensive such as BosWash, the US seaboard from Boston to Washington with 45 million people, and SanSan, coastal California from San Francisco to San Diego.

Megacities or **global hubs** are the key centres of the global economy: London (12.0 million people — 2010 estimate), New York (19.7 m) and Tokyo (32.4 m) are the centres of banking, finance and investment (the credit crunch may have diminished their reputations but they remain the key cities). In the future, Shanghai (16.6 m) and Singapore may have this status (but not population size in Singapore's case).

Millionaire cities are different from the global hubs in terms of population (2010 estimates): Mexico City (20.4 m), Seoul (20.5 m), Sao Paulo (18.8 m), Mumbai (19.7 m), Delhi (18.6 m) and Jakarta (18.9 m).

All characteristics become more extreme the higher the position in the hierarchy. A village might have a primary school and a pub while a city might have specialist cultural facilities, a university and a number of regional offices. These high-order functions give the city a very large sphere of influence. Obviously, the higher up the hierarchy the fewer the number of settlements.

Knowledge check 26

Define village and megacity.

Urban environments

Urbanisation is the process by which the proportion of the population living in an urban area increases. It is a process of change that affects both the places themselves and the people involved. A number of changes result as urbanisation occurs:

- A **sectoral shift** in the economy of a region or a country, moving from farming in the primary sector to manufacturing and the provision of centralised services which is linked to economic development.
- **Urbanism** — as people move to towns and cities either from rural areas or from other urban places, they change occupations as they strive to improve themselves. Urban dwellers encounter high costs of living and commute long distances. Inevitably, a substantial minority do not succeed and an **underclass** develops.
- The population **distribution** changes. A combination of rural to urban migration and the resultant natural increase from young migrants can result in rapid population increases in cities. Conversely, population loss from rural areas and an ageing population slows down the rate of rural population change. Some movement back to rural areas (**counter-urbanisation**) will also boost rural populations within commuting distance of urban areas.
- Changes within the settlement **hierarchy**. The number of urban areas grows rapidly as urbanisation occurs (e.g. nineteenth century UK). Small towns grow and larger ones coalesce to form conurbations.

Urbanisation over time

Knowledge check 27

Distinguish between urbanisation and urbanism.

There is a very close relationship between economic development and urbanisation. Figure 16 shows a model of the process that has been followed by MEDCs over time. (NB you only need to know the process as it applies to MEDCs.)

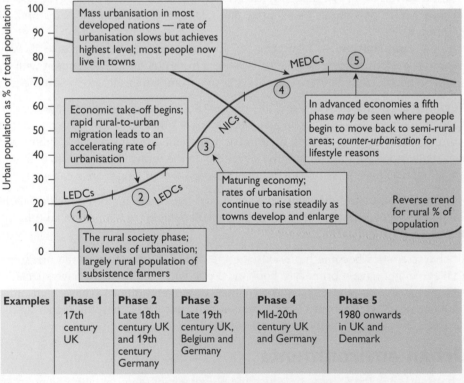

Examples	Phase 1	Phase 2	Phase 3	Phase 4	Phase 5
	17th century UK	Late 18th century UK and 19th century Germany	Late 19th century UK, Belgium and Germany	Mid-20th century UK and Germany	1980 onwards in UK and Denmark

Figure 16 The process of urbanisation

From approximately 1800 the UK experienced the Industrial Revolution, based on coalfields with extensive manufacturing industries. It took about 150 years for the urban percentage of the population to rise from 20% to 75%. Rates of urbanisation were slow because of high death rates with low life expectancy and very high infant mortality.

In the twentieth century, cities consolidated their hold on economies and developed a wide range of activities that employed an increasingly diverse population. By the end of the century, in advanced economies people were beginning to move back to semi-rural areas — **counter-urbanisation** — or back into cities — **re-urbanisation.**

The cycle of urbanisation

The present balance of centripetal and centrifugal forces shown in Figure 17 changes over time, either in response to commercial forces or in response to government planning decisions. The interplay of these forces has a major impact on the cycle of urbanisation and the evolution of the city structure and shape.

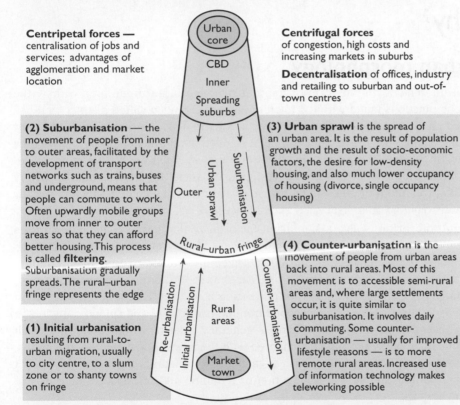

Centripetal forces — centralisation of jobs and services; advantages of agglomeration and market location

Centrifugal forces of congestion, high costs and increasing markets in suburbs

Decentralisation of offices, industry and retailing to suburban and out-of-town centres

(2) Suburbanisation — the movement of people from inner to outer areas, facilitated by the development of transport networks such as trains, buses and underground, means that people can commute to work. Often upwardly mobile groups move from inner to outer areas so that they can afford better housing. This process is called **filtering**. Suburbanisation gradually spreads. The rural–urban fringe represents the edge

(3) Urban sprawl is the spread of an urban area. It is the result of population growth and the result of socio-economic factors, the desire for low-density housing, and also much lower occupancy of housing (divorce, single occupancy housing)

(1) Initial urbanisation resulting from rural-to-urban migration, usually to city centre, to a slum zone or to shanty towns on fringe

(4) Counter-urbanisation is the movement of people from urban areas back into rural areas. Most of this movement is to accessible semi-rural areas and, where large settlements occur, it is quite similar to suburbanisation. It involves daily commuting. Some counter-urbanisation — usually for improved lifestyle reasons — is to more remote rural areas. Increased use of information technology makes teleworking possible

(5) Re-urbanisation involves a range of processes which enable people and economic activities to move back to city centres. **Gentrification** is when middle-class people move back to run-down inner urban areas and improve the housing stock. Some re-urbanisation results from planned initiatives such as those from Urban Development Corporations where inner central areas are improved in a number of ways, with high-value housing, hi-tech employment and improved environments

The cycle of urbanisation
- (1–5) is the typical sequence of processes in an MEDC.
- The processes result from the balance of centrifugal and centripetal forces. If there is outward movement, centrifugal forces are dominant.
- In some MEDC cities centrifugal forces have become so strong that planners are working to regenerate the centripetal forces.

Figure 17 The cycle of urbanisation

Examiner tip

Throughout this book you will see references to how people perceive places. In many answers perception of place is an important point.

Perceiving places

What people refer to as a village can vary. Most see a village in line with our definition but it is used particularly by estate agents to suggest a small community even if that community is engulfed in suburbs. Is Greenwich Village on Manhattan, New York a village or is it merely an area with distinct social and physical characteristics? Is the term 'market town' really describing the town with a rural market or is it trying to convey a small feel to a settlement that has expanded rapidly over the past 30 years?

2.2 How does the social and cultural structure of settlements vary and why?

Urban morphology

Urban morphology is the form or internal structure of an urban area or how the various functions found within a town or city are arranged. Often distinct land-use **zones** can be recognised. The main reason for zoning is the cost of land. **Bid rent theory** is often used to explain zoning. It states that all functions have a price they are willing to pay for a piece of land. For example, in the city centre, land prices are very high. Only land uses such as prestigious financial institutions and/or quality retailing can afford these costs (purchase price or rent and council tax) and so they are able to outbid other uses such as residences.

- **Agglomeration** is also a significant factor in zoning because many businesses benefit by grouping together within a business/office zone. Increasingly, positions at major road intersections are highly prized and valued by functions requiring maximum access to consumers — e.g. car showrooms, often called **motor rows**.
- **History** is a significant factor in influencing how a settlement developed. Urban areas in MEDCs generally developed from a historic core and spread outwards in a concentric fashion — so housing is generally older near the centre and much of it can be of poorer quality compared with the newer housing in outer suburbs. Sometimes historic buildings such as castles and cathedrals can distort the pattern by preventing spread in one direction or retaining an ecclesiastical quarter, e.g. Chichester.
- **Specialist site factors** influence morphology in a particular way. Many cities have a waterfront industrial area but the function has changed as the area became derelict. This has provided opportunities for **regeneration** (e.g. Cardiff Bay, Swansea SA1 and Albert Dock, Liverpool).
- **Political factors** (both central and local government and in the case of Wales, the Welsh Assembly) play a major part in how land is developed or redeveloped. City planning departments have strict laws on zoning — controlling where developments can take place. In 2011 the government was proposing to relax planning controls.

Urban areas are constantly changing. Figure 18 provides a checklist for the changing morphology of a city in an MEDC.

Examiner tip

You need to be able to draw a sketch map to show the morphology of at least one MEDC city (Figure 18).

Figure 18 Changes that may be occurring in an MEDC city

Urban models

An **urban model** is a simplified diagram created by looking at a range of cities and making generalisations, to include similar features. A combination of fieldwork (primary data collection) and map and census analysis (secondary research) can be used. The model can be compared to any city or town. You should try to explain why the morphology of your chosen city or town is similar or different. Always use an appropriate model for your chosen city. Some models such as Burgess' concentric rings were developed over 90 years ago for completely different circumstances (very rapid growth) from those experienced in modern MEDC cities. Other models, such as the multiple-nuclei, only apply to cities that developed from several centres, e.g. Los Angeles or Stoke-on-Trent. For UK cities, especially the northern industrial cities, Mann's model developed in the 1960s is useful (Figure 19).

A Middle-class sector
B Lower middle-class sector
C Working-class sector
   (and main sector of council estates)
D Industry and lowest working-class sector

1 Central business district
2 Transitional zone
3 Zone of small terraced houses in sectors C and D; larger housing in sector B; large old houses in sector A (Victorian)
4 Post-1918 residential areas, with post-1945 housing on the periphery
5 Commuter villages (post-1960)

Prevailing wind

Figure 19 Mann's urban structure model

Inequalities and social problems

Most urban areas consist of a mosaic of neighbourhoods of differing quality of life, but underclasses tend to be concentrated in the very worst physical environments of an urban area — usually because they have no choice. Richer inhabitants gravitate towards exclusive neighbourhoods so their paths do not cross with those of poorer inhabitants. Separation of rich and poor can become very marked, especially as the number of cyclical processes leads to a downward spiral of **multiple deprivation** in poor estates. And so problem areas develop (see Figure 20).

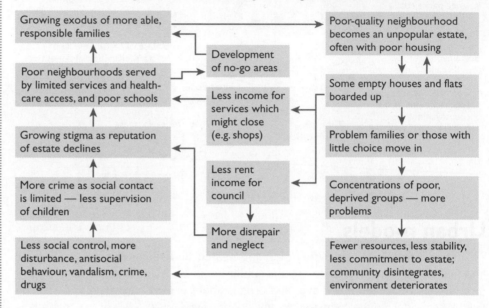

Figure 20 Spiral of multiple deprivation on poor estates

Cities are noted for their **multicultural plurality**, which can lead to a mixing of ethnicities, religions and cultures. It can lead to a mosaic of **ghettos**. A ghetto is an area where the population is almost exclusively made up of a single ethnic or cultural minority, e.g. Black and Puerto Rican Harlem in New York. The term comes from the area of Jewish segregation in Renaissance Venice. Ghettos are common in American cities and are increasingly found in European cities. They develop especially in times of unrest that lead to fear and the need to cluster. They also develop when new immigrants from different cultures enter a city, e.g. Turks in Stoke Newington, London. Ghettos are initially concentrations of deprivation. Many new immigrants arrive with limited resources, and discrimination can cause higher levels of unemployment and lower incomes among these groups. As ethnic groups are forced to concentrate in areas of poor housing, this pushes out groups from other races (called **white flight** in the USA). Levels of crime and unrest are frequently higher — hence the turbulence of places like south central Los Angeles. Figure 21 shows the factors that encourage ethnic segregation. In Milan, the Roma (Gypsy underclass of illegal and genuine immigrants) are housed in makeshift camps or under railway arches and are subject to forcible removal by the authorities — but they return.

Ethnic clustering is a complex topic and patterns of ethnic groups vary. In cities where immigrants are long established, well integrated, and not threatened by racial discrimination, the groups tend to disperse throughout the city as their confidence and affluence increases. This has happened in Sydney and Melbourne and is known as **mixing**. Multinational plurality prevails in many Australian cities.

Internal factors — those within ethnic groups that encourage segregation

New arrivals need mutual support from friends, relatives and community organisations

These immigrants will be happiest with religious centres, ethnic shops and foods, and banks grouping together to serve them

These immigrants need support from areas speaking their own language — a minority language in their new country

These ethnic groups encourage friendship and marriages and reduce contacts, except via schools, which might undermine the culture and traditions of the ethnic groups

Employment and accommodation can often be obtained via networking in an ethnic community

A closely knit ethnic community provides security against abuse and racist attacks — safety in numbers

Ethnic groupings help political power and influence development

External factors — those within the country or area that encourage ethnic segregation

As immigrants move in, the remaining majority population moves out — in fear of factors such as falling house prices

The majority population is generally hostile or unfriendly to new arrivals

Racism, abuse, racially motivated violence against ethnic minorities or fear of such actions

Discrimination in the job market — ethnic minorities in low-paid jobs or unemployed are forced by circumstances into cheap housing areas and substandard services

Discrimination by house sellers, estate agents and housing agencies keeps ethnic minorities in their ghettos

Discrimination by financial institutions forces ethnic minorities to use their own networks for small business development etc.

Ethnic segregation in urban areas leading to concentration

Figure 21 Factors that encourage ethnic segregation

- Some concentrations occur in the central area, such as the African-American ghetto in Watts in central Los Angeles. Sometimes this has been for historic reasons where there were large quantities of cheap housing to rent or to buy. In other cases, concentrations are in poor quality housing in the suburbs.
- Recent immigrants, such as Mexicans, have concentrated in **barrios** in, for example, east LA and San Fernando. As they secure better jobs and learn the language they may leave these barrios but will not do so if they feel racially threatened.
- Immigrants also live in what are called the **ethnoburbs** of the Pacific Rim: middle-class or professional immigrants settle in affluent suburbs where they invest in businesses and develop properties. These new clusters form due to choice, convenience and cultural reasons and are often influential economic enclaves.
- Elite 'ghettos' can occur in exclusive neighbourhoods such as the **gated communities** in Beverly Hills and in Esher, Surrey. Elite enclaves were common in colonial settlement where the colonial rulers were physically separated from the local people (sometimes these areas remain as enclaves, e.g. Victoria Peak in Hong Kong).

Most of the patterns outlined above result from natural processes, but under extreme racial or political tensions — as in parts of Northern Ireland or Bosnia — the fear and reality of persecution and extreme danger can lead to systematic **ethnic cleansing** or the withdrawal of various groups into increasingly 'pure' neighbourhoods, sometimes physically separated by a wall.

Knowledge check 29

What is white flight?

Ethnic concentrations are a major cause for concern when ethnic minorities become an underclass concentrated in the least attractive, socially excluded estates. In MEDC cities these are commonly rundown inner-city areas, but they can equally be problematic estates of **social** or **welfare** housing either in the inner city (Hulme in Manchester) or on the city outskirts (the Parisian *banlieues* such as Grigny, La Courneuve and Sarcelles).

The changing employment market has affected both types of area, with many low-paid, part-time, insecure jobs available together with high levels of unemployment, especially among young adults. The cycle of disadvantage and deprivation is very difficult to break as it tends to be passed from generation to generation — a fact reinforced by poor health and educational facilities in these areas. It can also be reinforced by racial and other forms of discrimination.

Solutions

A number of controversies exist as to how to best solve problems of inequality:

- Should the finance come from public or private sources or a mix of both? The 2008/09 credit crunch had its origins in private finance for loans in poor areas that led to toxic debt.
- Should the initiatives come from central government (top down) or via **localism**, local community-based projects?
- Is tackling the environmental, the social, the economic or the political problems the key priority?
- What is the best way to spend any money allocated — saturation in one concentrated area or dispersed initiatives?

Most cities have their own websites detailing their schemes.

Studentification

A more recent phenomenon of segregation is the development of student districts in university towns. Once these were the colleges of Oxford and Cambridge, but today large areas of inner-city wards near universities have become residential districts for students, e.g. Southsea for the University of Portsmouth.

Perceptions of urban structure

Most people carry a mental map of an urban area based on their experience and education, as well as the further experiences they receive as they journey about the area. Lynch modelled the way people see cities:

- **Paths** — our channels of movement along roads or rail lines or where we walk or what we will see as we use the images in Streetview or from a sat nav.
- **Edges** — the barriers that cut areas off from one another, such as urban motorways, rail lines, a river, coastline, or even the boundary of a gated community. Some edges do become **seams** that bind areas together, such as the Cam in Cambridge, which binds the colleges into a whole.
- **Nodes** — the focal points in a town or city where major transport pathways meet or where there is a concentration of activity. Railway stations are nodes in most people's minds. Canary Wharf is a node as is the Welsh Assembly building in Cardiff.

Examiner tip

Always check the bias of the sources — the official statements on the success of improvement schemes are often more optimistic than the views of the people living there.

Knowledge check 30

Why might students live in one area of a city?

- **Districts** — areas identified by a common characteristic, normally its architecture. Georgian Bath is an easily perceived district as are many city centres, although these may be perceived in part rather than as a totality.
- **Landmarks** — the features with distinctive characteristics that people remember. They can be very local, distant landmarks or even outside the urban area.

How people see all of the above depends on:

- Their age — a 3-year-old child may see the 'dong clock' — Big Ben — but not notice the Thames. Young adults might see more in an entertainment district than their parents who perceive it differently.
- Gender.
- Experience and frequency of visiting an environment — familiarity aids recall.
- The social class of the individual or group — for example, are 'no-go' areas perceived as a result of social class or perhaps they are areas where housing is not for sale and only available for rent.
- The value system of the individual or group. Many ethnic and religious districts grew up because they were the areas where a person's value system was best appreciated, e.g. the Jewish area around Golders Green and South Koreans in Merton.

2.3 What are the issues of the inner city?

Inequalities

Inequalities are found in all urban areas and the scale of the inequalities reflects national and regional patterns. Urban models show that contrasts in wealth can be found over very short distances. Both the wealthy and the poor concentrate spatially and there are reasons for this **social segregation** in the inner city and beyond.

The key is housing, which is fixed in location. Developers and planners tend to build housing with a particular market in mind. Wealthier groups can choose where they live — paying premium prices for areas well away from poor areas, with pleasing environments and quality services, such as schools. The poorer groups have fewer choices and are, in a sense, forced to live where they are placed — in social housing or where they can find a cheap place to rent privately.

Housing needs to be considered in the context of changing environments. Housing neighbourhoods change over time — large Victorian villas are now too large for the average family and so have been 'converted' into apartments for private renting. Other neighbourhoods have improved because of **gentrification**. The same type of housing can be built in different environments. It may be on an attractive hillside (Hampstead) or in a recently improved neighbourhood targeting upwardly mobile people. **Right to buy** transformed some council estates so that, for example, rows of houses facing fields were bought and improved. As we have seen, immigration can result in inequalities.

Measuring inequalities

It is possible to measure **quality of life** in an area, such as the inner city, using **primary** data. This data includes quality, density, and condition of housing, and the nature of the environment (physical and social). Levels of pollution or crime are good indicators here (Figure 22).

Physical
Quality of housing; levels of pollution in environment (air/noise); incidence of litter, graffiti, vandalism

Political
Opportunities to participate in community life and to influence decisions (percentage voting in elections)

Quality-of-life measures

Social
Incidence of crime; fear of crime; standards of health and access to health facilities (percentage in poor health); standards of education (percentage staying on post-16, percentage educated to degree standard etc.); percentage on free school meals and on state benefits (e.g. disability)

Economic
Access to services, range of choice and quality; access to leisure services, open spaces etc.; access to employment (percentage unemployed long term, youth unemployment); levels of income; percentage of lone parent families; percentage of lone pensioners

Figure 22 Quality-of-life measures

It is possible to use **secondary** data from a census to assess **deprivation** levels — this may include **poverty** in terms of low income or can be manifested in poor health or lack of possessions, e.g. a car. It is very common for the poorest parts of an urban area to suffer from **multiple deprivation** — with social, environmental and economic deprivation. There have been many indices devised to measure deprivation, including Townsend's.

Urban social exclusion is a term that refers to the problems faced by residents in areas of multiple deprivation, such as the inner city, where people are excluded from full participation in society by their social and physical circumstances. They cannot get access to a decent job because of poor education, or obtain decent housing because of poverty, and have to suffer high levels of crime and poor health in an unattractive physical environment.

Examiner tip
Make sure that you have your own examples of areas of deprivation and poverty in cities, especially inner cities. However, do not neglect poverty in the suburbs.

Figure 23 Idealised quality-of-life transect across a UK city

Inequality can have major consequences in a city in terms of a lack of social cohesion. In extreme cases it can lead to civil unrest as may have been the case in the 2011 riots. Governments have to address social injustices for a variety of social, economic and political reasons. Figure 23 shows an idealised quality of life transect across a city in the UK and the lower quality of life in the inner city.

Urban issues are summarised in Table 8. Many of these are wider in extent than just for the inner city although they are measurably more intense in the older parts of a city.

Table 8 An overview of urban issues

Environmental	• Problems caused by atmospheric pollution from a variety of sources, especially from industrial, domestic and transport services • Unsustainability of the energy used in commuting • Problems of traffic congestion — moving the daily flow of commuters and overloaded public transport • Road accidents leading to death and injury • Problems caused by sewage and waste disposal • Issues of water pollution • Noise pollution • Problems generated by rapid urban sprawl, in particular loss of land and creation of wide areas of slums • Dereliction may be a problem, with many 'toxic' or unsightly areas such as former gas works • Some cities have specialist problems such as flooding or landslides — particularly problems for poor people who cannot choose to live in safe locations
Social–cultural	• The development of zones of poverty and dependence within the city — can become areas of high crime and social deprivation • Ghettoisation of various income and ethnic groups is leading to friction between groups/gangs • Creation of an underclass of people with limited powers to challenge their poor quality of life and health and dependent on state assistance • Drug use and crimes resulting from gang warfare
Economic	• The problems of providing for the enormous numbers of new arrivals • Services such as health, education, housing, employment and utilities such as water and sewage are increasingly costly to provide • Costs in terms of well-being
Political	• The difficulties of governing mega-cities effectively to cope with the enormous range of problems; declining incomes from council taxes due to capping

Housing stock

Housing problems are some of the most widespread and serious of all the troubles facing inner cities. They have many facets:

- The quality of housing — size, quality of construction, provision of services such as water, electricity and gas, and the safety of the site. Poorly constructed houses that are overcrowded have a major impact on the health and well-being of residents. For example, consider many of the problems associated with high-rise blocks in the inner city.
- The quantity of housing relative to the number of households that require homes. Often the demand for housing exceeds supply. This has a knock-on effect on other factors, such as finding workers. Where there is a shortage of housing, some people are inevitably housed inappropriately.
- The availability and accessibility of housing — the ease with which people are able to buy or rent or self-build housing at affordable prices. The right to buy policy

removed large quantities of inner-city housing (and outer city estates) from the renting sector. In many cities prices are rising and people are unable to get onto the private housing ladder. This is why affordable social housing in MEDC cities is of vital importance in solving inner city housing problems.

- The range of housing tenure — the housing mix. A good balance between the supply of owner occupied, privately rented and social housing in relation to demand is essential in all cities.
- Increases in homelessness caused by unemployment, defaulting on mortgages and family break up.

Knowledge check 33

What is meant by multiple deprivation?

The downward spiral of housing for the urban poor

While the rich have the opportunity to choose where to live, providing decent housing for the poorer population of urban societies is a major challenge. Low incomes mean little or no choice — being forced to rent, with little security of tenure or protection against sudden or forced eviction. Poor people have to put up with inadequate housing, for instance in tower blocks or overcrowded apartments. The houses are normally in unsafe locations, often perceived as problem areas, which prevent their inhabitants from access to credit. In many cities, in spite of attempts by national and local governments, the problems are getting worse rather than better because decent, well-located yet affordable housing is in very short supply. Many people have to travel long distances to work in order to obtain reasonable housing. As can be seen in Figure 24, there can be many areas of poor housing that are not located in the inner city.

Usual solutions
(1) • Inner-city regeneration
  • Slum clearance
  • Down-topping of high-rise flats
  • Housing action areas
(2) City challenge or SRB schemes for all areas including outer-city council areas
(3) Gentrification of some areas, not in itself a solution to housing problems but does improve the environment
(4) Re-urbanisation of derelict areas to provide more high-value, quality areas

(1) Inner-city areas, a mixture of redeveloped nineteenth-century slums and modern high-rise flats
(2) Inner-city transitional areas; former villas have become bedsits and multi-let homes
(3) Outer-city council estates, many of which have become very deprived

Figure 24 Areas of poor housing in an MEDC city

One response to the strategy of developing appropriate housing within the inner city is **gentrification** (Figure 25). This is the process by which relatively affluent middle-class households move into desirable inner-city neighbourhoods. It applies to neighbourhoods, but often has knock-on effects and steadily spreads. Some planned gentrification of old warehouses in docklands is on a very large scale, e.g. the Wapping area of London Docklands.

Gentrification in the balance

+
- Investment improves the environment of run-down areas
- Housing is refurbished and improved — it could have become derelict
- Rich people move into the area and their spending brings new businesses (e.g. wine bars) and increases income from local taxes, community charges etc.
- The new dynamism represents an opportunity for development
- It is paid for by individuals, not the council — although grants may be used

−
- Gentrifiers take 'out of commission' housing that poor people could have lived in
- Gentrification leads to rising house prices and rents, thus reducing housing supply
- Some gentrifiers are 'developers' for profit
- New 'yuppies/dinkies' cause friction with traditional inner-city people
- Many gentrifiers are car owners in spite of moving to cut commuting — car parking in streets becomes a major problem
- Many traditional local services, such as schools and corner shops, are not used by these arrivals
- Can be very expensive to restore 'historic' houses and other properties

Figure 25 Gentrification in the balance

Gentrification is a component of the **re-urbanisation** process that is taking place in the inner city to combat urban sprawl and overdevelopment of **greenfield** sites. Another strategy is to build new-style housing — mainly for young executives — on **brownfield** sites such as former hospitals (e.g. Graylingwell, Chichester) or closed teacher training colleges (e.g. Park Centrale, Southampton), or to refurbish old industrial buildings to provide luxury apartments. Sometimes, as in London Docklands or Newcastle-upon-Tyne, inner-city nineteenth century terraced housing has been knocked down displacing the existing population. Here, large-scale development of upmarket housing is seen as part of the overall process of regeneration and environmental improvement. The displacement of poorer people from large parts of the inner city, when in fact they form a vital component of the workforce that keeps a city running, is a major issue in MEDC cities. Displacement, combined with changed family circumstances, can generate the first stage in the cycle of homelessness. A further aspect of regeneration is the conversion of empty offices for residential use, which was being proposed in the City of London in 2011.

Knowledge check 34

What is brownfield site regeneration?

Regenerating and redeveloping urban areas

Figure 26 summarises the major issues of regeneration.

Regeneration can lead to a number of possible improvements:
- Dealing with social problems such as poor health, low educational standards or substandard housing in **action zones**. These programmes were community based, although they could be top-down in design.
- Targeting economic improvements that relieve poverty and thus solve some social problems or provide new employment, e.g. **urban tourism** in Bradford.
- Improving the environment, for example turning derelict waterfront properties into an attractive environment provides a catalyst for development. This has been particularly successful in Barcelona and many other former dockland areas.

- **Image enhancement** or **urban boosterism** or **rebranding** — by promotional programmes to change outsiders' perceptions (vital to attract inward investment). Often based around promoting cultural and sporting facilities, e.g. the 2012 Olympic sailing events in Weymouth.
- **Prestige project developments** (known as **flagship projects**) — the 2012 London Olympic Park (including the athletes' village and the Westfield Stratford City Shopping centre), waterfront developments (Cardiff Bay and the Welsh Assembly building) and convention centres (Albert Dock, Liverpool). The major issue here is to what extent these projects directly benefit the people of the area. Also, much money can be wasted if projects are unsuccessful (e.g. the O2 arena for several years after 2000).

Knowledge check 35

What do you understand by boosterism?

A number of governments have consistently tried to regenerate declining urban areas. The British case study below summarises the stages and the schemes. Such schemes reflect the political orientation of the government in power.

Environment
- Redeveloping extensive areas of derelict land, largely in the zone in transition around the CBD and in minor urban areas
- Cleaning up land contamination and water pollution which are barriers to development

Government
- Big issues — increasing disenchantment of electorate
- Disenfranchising, disinterest, detachment of urban poor

The problems

Socio-cultural
- Cities are increasingly divided, with polarisation of rich and poor
- Ghettos and areas of social exclusion are almost impossible to regenerate, even with a huge investment

Economic
- Massive decrease in the number of jobs because of deindustrialisation
- Centrifugal forces have encouraged industry into semi-rural greenfield sites
- Consequences of very high unemployment or under-employment in low-paid jobs

Image
- Many former industrial cities have a negative image of dereliction, dirt, danger from crime, decay and deprivation — not places to invest in

Figure 26 The major issues of regeneration

Phases of urban regeneration in the UK

Phase 1: post-war recovery — focus on housing 1947–67

The main aim was to provide people with housing. Many cities were heavily bombed.
- **Comprehensive Redevelopment Area (CDA)** programmes demolished slum areas and re-housed people in high-rise blocks (tower blocks). At the same time, utilitarian council estates were built in the outer suburbs (e.g. Leigh Park and Paulsgrove, Portsmouth). Some housing was erected as prefabs — emergency homes.
- **New Towns** — not inner-city but a part of the process that removed population from the inner city, e.g. Cwmbran and Crawley. Also, **Expanded Towns**, e.g. Andover, were designed for the overspill from urban areas in what were seen at the time as quality environments.

Despite these efforts housing shortages were still widespread and the dwellings were of poor quality. Redevelopment did not keep pace with decay and slum clearance led to areas of dereliction.

Phase 2: social and economic action 1967–77

Attention began to focus on the inner-city zones that had experienced **decentralisation** and **deindustrialisation**. The arrival of immigrants into, for example, Brixton (London), Chapeltown (Leeds) and Moss Side (Manchester), compounded some of the problems.

- In 1968, the **Urban Aid Programme** established the idea of government grants for a range of small-scale projects in deprived areas: housing renovation, advice centres and nursery schools, as well as community development projects.
- In 1969, **General Improvement Areas (GIAs)** were established and in 1974 **Housing Action Areas (HAA)**.
- **Education Priority Areas (EPAs)** were established in 1967 followed by **Community Action Programmes (CAPs)** in 1974 and then a host of other small initiatives.

There were many ideas, but little coherence and insufficient funding and it was difficult to see lasting, tangible benefits.

Phase 3: Inner-city focus 1977–90

The government recognised the inner city problems and focused efforts on economic regeneration, environmental improvement and creating services for communities.

- **Partnership Areas** (e.g. Salford), which received annual central government funding for 23 programmes including the Govan area of Glasgow.
- **Urban Development Corporations (UDCs)** were introduced in 1981 as a property-led development to regenerate large tracts of land (e.g. London Docklands and Cardiff Bay). UDCs had similar powers to New Town authorities and were answerable to central government. Issues included: job creation, over-reliance on property development (which was only beneficial in times of rising prices), reduction in the power of local government, and public money used to encourage private development.
- **Enterprise Zones (EZs)** were introduced in 1981. These covered small areas of under 450 hectares and were expected to offer special incentives to attract high-tech industries by relaxing planning regulations. They were sited in areas of high unemployment, e.g. Gateshead and the lower Don Valley, Sheffield. Doubts were cast on the number of new jobs created because jobs migrated into the zones depleting the surrounding areas of jobs. Many zones failed to attract innovation.
- **Task forces** were small scale teams that acted as think tanks to saturate a deprived area with aid and grants (e.g. St Paul's, Bristol, and Moss Side, Manchester).

Phase 4: Approaching the Millennium 1990–2009

This phase was associated with competitive bidding — the best projects got the money.

- **City Challenge 1991** involved private, public and voluntary bodies working together. The areas involved had high unemployment, low skills base, environmental dereliction and deteriorating housing (Wolverhampton) and housing improvements (Hulme, Manchester). While there were many fine projects with good private sector investment and imaginative ideas, the competitive situation wasted energy and time.

Unit G2: Changing Human Environments

- **Single Regeneration Budget (SRB)** was introduced in 1994 to join up all the sources of funding to make a coherent regeneration project. Cobridge, Stoke-on-Trent (containing the red light district) was a good inner-city scheme.
- **Millennium schemes** like the Millennium Stadium in Cardiff and the Millennium Dome (now the O2 stadium) at Greenwich had a mixed reaction.

Phase 5: The retreat from state intervention 2010–

- Eleven English **Enterprise Zones** as part of the 2011 Plan for Growth in Manchester, Birmingham, Merseyside and Newcastle, applied for by Local Enterprise Partnerships were economic measures to solve urban problems through private investment tax breaks and job creation.
- Second wave of 11 **Enterprise Zones** using old military land, e.g. the Solent Enterprise Zone at Daedalus Airfield in Gosport and Alconbury Airfield, near Huntingdon in Cambridgeshire, focusing on science-based industries, e.g. Daresbury Science Campus in Warrington.
- Simplifying the planning system, with a presumption that all development should be sustainable and inherently beneficial, in order to increase the number of houses available and meet demand.

Three trends emerge from this case study:

(1) The increasing role of central government in the inner city, followed by...

(2) Increased private investment usually following government 'pump priming' or 'matched funding'.

(3) Increased movement away from social schemes, with economic consensus seen as the key to improvement.

Knowledge check 36

What are Enterprise Zones?

When evaluating urban regeneration there are a number of criteria to use, particularly: were the objectives of any particular scheme met, and were the targets reached within the allocated period of funding?

Figure 27 sets out an evaluation framework. You should use this for your chosen urban regeneration scheme (e.g. Cardiff Bay, Swansea SA1). While there have been successes — cities such as Glasgow and Birmingham have transformed their images — a number of concerns have been expressed about regeneration in other cities, especially those carried out by the Urban Development Corporations (e.g. London Docklands).

Examiner tip

Look up where the most recent enterprise zones have been located. Why are they in these places?

Percentage achievement of targets, e.g. amount of land reclaimed or number of houses built

Costs versus benefits — amount of money invested and any problems generated

Tangible products and success of function, e.g. number of tourists

Net number of permanent jobs produced and at what cost

Environmental quality before and after — pollution monitoring etc.

Ability of schemes to help urban poor, e.g. percentage of new social housing; opinions of original inhabitants

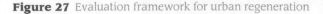

Figure 27 Evaluation framework for urban regeneration

Solving inner city traffic problems

The following are solutions to the inner city traffic problems listed in Table 8:

- Build and redesign roads and traffic schemes — not very practicable in the inner city because it cuts off areas from one another.
- Promote public transport — trams, bus lanes.
- Traffic management including traffic calming measures to stop rat runs through the inner-city streets and congestion zone charging.
- Discourage use of private cars and discourage no-charge parking areas in the inner city.
- Make cities more compact including the inner area.
- Make cities more cycle friendly.

Knowledge check 37

What were Comprehensive Development Areas and Expanded Towns?

2.4 What are the issues in the CBD?

The nature of the CBD

CBDs become more distinct the larger the city. A variety of zones can develop within the CBD, a product of the changes taking place in society and the economy, and a response of city governments to those changes. The key zones of the European city centre are:

- An old centre that was the medieval core of the city, which contains the cathedral or major church and old coaching inns, the old street market and some recreational functions such as restaurants.
- The modern retail core which may have developed out of the historic core. This is characterised by major national chain stores that have caused people to refer to High Streets as **clone towns** because the image presented to the public by each store is the same in every town. In the core of the retail region the function will cover all floors. It will also be the area of greatest pedestrian flows (**footfall**). In some cities there may be conserved aspects of former retail architecture, e.g. covered arcades as in Cardiff and Leeds. Often the area might contain a modern multi-storey centre, e.g. St David's 2 in Cardiff, and the Arndale in Manchester.
- In some cities there is a smaller area of specialised retail outlets that cannot afford the high rents of the inner core.
- **Quasi-retailing** is the name for banks, building societies and other financial outlets that in medium sized cities can be found in a separate area.
- The office district has grown to be a separate part of the modern CBD. The larger the city population, the greater the scale of this main employment focus. In the largest cities the office area can have regions within it, for example, the legal profession may be in a distinct area, possibly near the courts. Separate stockbroking areas and insurance districts may develop, e.g. City of London.
- Some towns and cities have a governmental area where council offices, the police and law courts might be found. In capital cities this will form a large swathe of the central area such as Westminster, Whitehall and Victoria Street in London and the administrative offices clustered north of the centre in Cardiff.
- A growing feature of the modern CBD is a recreational district of restaurants, bars and clubs. By day these areas are functioning to support the CBD but by night they can become no-go areas for many people.

Examiner tip

Make sure that you keep discussion of any CBD issue to the actual CBD unless the question permits you to look at the zone in transition.

Growth can be constrained by areas such as a Victorian park and rivers act as natural barriers to growth except in the largest of cities. Rail lines can also constrict growth. Some of these areas may not be present in CBDs in North America and other parts of the world.

The CBD is surrounded by a **Zone in Transition or Frame** between it and the inner city (Figure 28).

The changing CBD

Most CBDs in cities in MEDCs are currently under threat for six main reasons:

(1) The loss of retailing functions to out-of-town shopping centres and **retail parks** and the decline of the traditional High Street.

(2) The loss of offices to suburban or peripheral locations in prestige **science parks**.

(3) The nature of the CBD itself, which because of congestion is becoming increasingly inaccessible, especially as the bulk of the population lives in suburban areas. Equally, the CBD has very high rents and council taxes and is only affordable to prestige businesses, so centrifugal forces begin to dominate.

(4) The effects of the 2008/09 credit crunch with the closure of retail chains, e.g. Woolworths in the UK, and the shrinking of the financial sector.

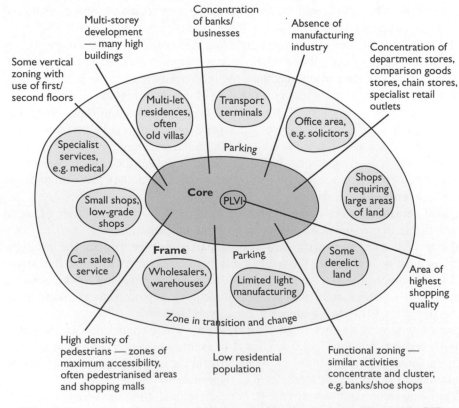

PLVI = Peak land value intersection — the highest rated, busiest, most accessible part of a CBD

Figure 28 Key features of the CBD core and frame

(5) The growth of central recreation areas including museums, theatres and galleries.

(6) The rise in popularity of internet shopping, reducing the number of visits to central areas.

Figure 29 Factors influencing CBD decline

Reviving city centres

A number of strategies have been devised to enable city centres to remain vibrant:

- Using a city business and marketing team to coordinate the management of the city centre and encourage special events.
- Pedestrianisation to improve safety and to provide a more attractive shopping environment with new street furniture, floral displays and landscaping.
- Encouraging the construction of all-weather shopping malls/centres (air-conditioned in summer and heated in winter) in key locations in the city centre, often with integral parking.
- Encouraging the development of specialist areas, including attractive open street markets, cultural quarters and specialist arcades selling high-quality products such as organic foods and local handicrafts.
- Improving public transport links with rapid transit into the central, pedestrianised area, e.g. park-and-ride, shopper buses.
- Planning car parks to give ease of access to the centre whilst also controlling pricing and time parked.
- Making the centre safer with video surveillance.
- Developing new leisure and cultural facilities.
- Promoting tourism activities to encourage greater spending by **conserving the heritage** — used in Bath, York, Chester and Cambridge to great effect.

In the USA, where shopping malls are widespread, the movement to reinstate and revitalise the town centre is strong as both planners and the general public realise the dangers of dead city hearts (**urban doughnut effect**).

Round-the-clock city centres

British city centres are frequently 'dead' and dangerous at night. By encouraging shoppers to stay longer and perhaps dine out, they may be more vibrant. Here are some possible schemes:

- Run shopping events such as farmers' markets and Christmas Fairs.
- Late night shopping and Sunday opening.
- New retail developments focused around a major 'magnet' store, e.g. West Quay, Southampton, with John Lewis as the focus and IKEA nearby.
- Plan for a wider range of leisure facilities such as cafe bars, music venues, theatres and cinemas that people would naturally visit during an evening. Some of these will be located on the CBD fringe.
- Promote street activity by permitting cafes to spill onto the street (in the UK, only realistic in summer).
- Develop nightlife and clubbing — although the costs may exceed the benefits.
- Plan for theme areas such as the 'Gaybourhood' area of Philadelphia or the cultural areas of Stoke and Sheffield.
- Bring back residences into the dead heart — flats to rent above shops or the development of old buildings as flats (a strand of gentrification) targeted at higher-income workers.

Most city managers are trying a range of these strategies and shoppers are being attracted back to centres. These strategies can take place in conjunction with controlling permission for and limiting the number of suburban (e.g. the 1.9 m square feet Westfield Stratford City Mall, Newham), out-of-town shopping centres and giant super-malls such as the Trafford Centre, Manchester.

Studies of the impact of super-malls on traditional town centres conclude that smaller towns are most affected. For example, Merry Hill in the West Midlands has affected Dudley far more than it has Birmingham.

Tackling Question 3c through CBD study

Question 3c in your examination will expect you to be able to write about **your own study of a topic in human geography**. You will be expected to gather information, know sources of information, have acquired the skills of mapping and field observation, interpret information realising that there may be bias, and draw conclusions from that information. The CBD provides the ideal opportunity for you to develop these skills through a field exercise.

CBD enquiries that you could attempt

(1) An analysis of retailing in the CBD (e.g. Shrewsbury).

(2) The distribution of one retail type in a large CBD (e.g. Cardiff).

(3) What shops have closed and where are the vacant shops located? Newport had 26% of shops vacant in September 2011 so would be a good case study.

(4) Pedestrian flows and retail types.

Examiner tip

Your approach to this part of your study of Changing Human Environments should follow the sequence of enquiry that might be modified by your teacher.

(5) Street markets and pedestrian flows when a market is in operation and when not in operation.

(6) Differences in retailing between pedestrian streets and streets with traffic, or between arcades, streets and covered centres (e.g. in Cardiff or Leeds).

(7) Analysis of distribution of outlets in a new covered centre (e.g. St David's Centre or Westfield Mall).

(8) The distribution of banks, building societies and other financial services.

(9) The changing nature of leisure in the CBD.

(10) Day and night leisure.

(11) The nature of the CBD frame.

(12) Is there a zone of discard?

(13) Is the centre a 'clone town'?

Examiner tip

Topics 9, 10, 11 and 12 should be studied as a group for safety reasons.

The route to enquiry

(a) Plan the enquiry

(1) Read about the CBD in textbooks and journals.

(2) Decide on a topic that is either an area of investigation (e.g. defining the edge of the CBD), or a question that you wish to answer (e.g. do shops cluster in the CBD?).

(3) Plan how you will get the information and data from primary (fieldwork/questionnaires) and secondary sources (Goad plans) that will enable you to answer your research topic. Work out the categories of shops that you wish to map. Prepare data collection sheets. Do you need others to help you?

Examiner tip

Make sure you have proper maps. Rough sketches do not enable you to be thorough in your analysis.

(b) Carry out the fieldwork/collect the data

(1) Record your data on data sheets and/or maps. Know what sampling procedure you used.

(2) Know the faults or errors in your data collection methods.

Examiner tip

You could be asked about faults or errors in your data collection methods in the exam.

(c) Process the information

(1) Map the distributions. Understand what the information that you have gathered actually says about the clustering of shop types that you are studying. How accurate is that information? Is it biased? If so, why?

(2) Try to process data statistically.

Examiner tip

You will not be asked about the statistical processing in the examination but you need to know what the statistical processing shows.

(d) Write 1000 words on your study

(1) Organise the information and data so that it answers your question, for example: Which shops cluster — shoe shops? Do large stores cluster? Why do some cluster while others are spread out?

(2) Draw conclusions from your results.

Examiner tip

Always know your conclusions because they will convince the examiner that you know what you are talking about.

(e) Evaluate your work

(1) Evaluate each stage of your study so that you can answer questions about your sequence of enquiry in the examination.

(2) Evaluate your evidence and conclusions in relation to what you read at the outset and what you have found. This should include altering the topic or the question that you asked so that others could gain a better understanding of the topic in the future.

Questions that could be asked on your study

Stage (a)

(1) Outline how you planned your study into a changing human environment.

(2) Outline the planning stages of an investigation that you have undertaken into a changing human environment.

Stage (b)

(1) Describe the methods that you used to gather data and information for an investigation that you have undertaken into changing human environments.

(2) Evaluate the methods of data collection that you used for an investigation that you have undertaken into changing human environments.

Stage (c)

(1) Evaluate the methods used to present information in your investigation.

(2) To what extent and why was your information biased that you gathered for an investigation that you have undertaken into changing human environments?

Stage (d)

(1) For an investigation that you have undertaken into changing human environments, outline the conclusions that you made to your study.

(2) Discuss the main findings of an investigation that you have undertaken into changing human environments.

Stage (e)

(1) To what extent is it possible to increase the validity of the conclusion of the investigation that you have undertaken into changing human environments?

(2) Attempt an evaluation of your sequence of enquiry for an investigation that you have undertaken into changing human environments.

All these questions could be answered if you carried out an investigation on the CBD.

Knowledge check 39

What does the command 'evaluate' mean?

Examiner tip

This sequence should be used whatever your study. See if your study has addressed all of these stages.

2.5 How is the rural–urban fringe changing and why?

The rural–urban fringe is very much a zone in transition from the built up area to the countryside — a highly desirable area for which many land uses compete (Figure 30). Urban areas gradually extend into the rural–urban fringe and cause many pressures and conflicts.

(1) The **urban fringe** is a zone of advancing suburbanisation. High-quality houses on the fringe of an urban area are always in demand — the urban fringe is perceived as quieter, less polluted, more crime-free and less congested, often with access to quality services such as good schools. With perceived beautiful views over countryside, living at the urban edge is seen as combining the advantages of town and country living. In general, the homes built here are more highly priced and sell more easily. But as the urban fringe is always advancing (unless it is controlled by a green belt), the rural 'idyll' may become, after a few years, an illusion as the open view disappears and land is gradually released for development or ring roads.

(2) The **green belt** is used to control urban sprawl. It is an area of open farmland and recreation land around an urban area (mineral extraction is permitted). The only development that can take place in green belt areas is within existing settlements (**infill**), or where planners decide that local, regional and national needs will be met by an extension to the settlement. The Cambridge Science Park was a case in point because it brought new jobs to the region. In theory, by restricting growth in the green belt, developers are being encouraged to use **brownfield** sites in inner urban areas. The reality is that because of the higher cost of developing brownfield sites, the developers merely start building beyond the green belt in the commuter belt, a process sometimes called **leapfrogging**. There is evidence that green belts are being reduced in size, but they are also very restrictive. Some planning authorities form **green wedges** or **green corridors**, e.g. between the Ruhr towns in Germany, because they allow more development yet also conserve open countryside. Because of their rarity value, any developments in the green belt can command very high prices and housing that pre-dates the creation of a green belt soon becomes exclusive.

(3) The **commuter belt** has developed beyond the green belt in response to demand. Where there is a motorway or fast road, large dormitory settlements develop — so called because commuters only come home to sleep there. Technically, as these occur in semi-rural areas, they are the products of the process of **counter-urbanisation**. But these settlements are very suburban with large estates of private housing, out-of-town supermarkets and leisure facilities. However, in the sectors between the main transport routes, commuter developments are more restricted — often to small estates of under 10 houses, lifestyle hobby farms, barn conversions and renovated country cottages. Counter-urbanisation is also the product of firms moving to small towns and taking their workforce with them. Some people also move to this zone for retirement but are dependent on housing where there are basic village services.

(4) Beyond these zones are the beginnings of a rural zone that is used by city dwellers for weekend recreation. Clearly the areas under most pressure are the attractive and accessible areas. **Country parks** have been developed in this zone specifically for day trippers. The original idea of these country parks was to take pressure off high value, environmentally sensitive sites in National Parks.

Knowledge check 40

Distinguish between a green belt, a green wedge and a green corridor.

Examiner tip

Do not become too emotive and fall into the trap of making all rural areas conform to the rural idyll. Write out your own categorisation of a range of villages in your area.

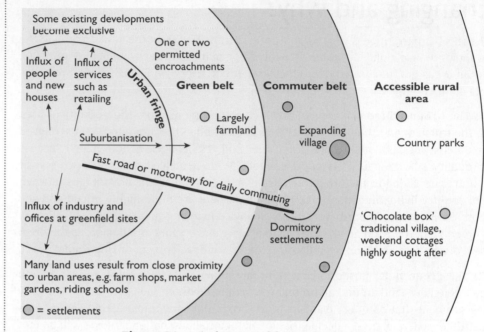

Figure 30 Main features of the rural–urban fringe

Types of development in the fringe

- Intensive and/or agribusinesses such as market gardens and pick your own.
- Playing fields used by urban dwellers for formal and informal recreation.
- Golf courses, shooting ranges — extensive recreation.
- Park and ride terminuses (e.g. Oxford).
- Science and Technology Parks (e.g. Cribbs Causeway, Bristol).
- Hotels like Travelodge and Holiday Inn Express.
- Major sports stadia (e.g. Falmer Stadium, Brighton).
- Industrial sites (e.g. Rolls Royce in a disused gravel pit near Chichester).
- Retail and office parks (e.g. Hedge End, Southampton).

The rural–urban fringe is a mosaic of land use and land types — from high-quality to degraded or derelict land — all within a small area and influenced by four main factors: agricultural policies; the nature of the agricultural economy; the urban economy; countryside planning regulations.

Many land uses look attractive but others can be ugly. Because the area is under pressure from competing uses, its state and appearance can change rapidly. Also, many land uses, like golf courses and sailing in former gravel pits, are not very natural and are controversial. Conflicts result from:

- **Commercial decentralisation** — industry seeking rural greenfield sites as opposed to city brownfield sites.
- **Housing demands** place pressure on the rural fringe — a strong desire to build in green belt areas and attractive villages.

- **Farming** — problems of trespass, vandalism, fly tipping and dogs, where people seeking recreation try to use farmland.
- **Environmental issues** — much of the development of roads, new science parks and new housing can cause increases in pollution and also heighten flood risks.

Here are some potential investigations in the rural–urban fringe:

- Retail parks and superstores and their impact on the environment and other retail outlets.
- Leisure centres and new parks on reclaimed land such as former gravel pits.
- Marinas in reclaimed areas.
- Loss of school playing fields.
- Office and science parks and impact on commuting and employment.
- Problems of intrusions onto rural land — fly tipping, horticulture.
- Former airfields now engulfed by urban areas (e.g. Brooklands, Surrey).
- The impact of leisure developments such as golf courses, paint balling, mountain biking, motocross, leisure centres and rowing/sailing lakes.

Examiner tip

Threats to and issues in the rural–urban fringe such as those above can form a worthwhile investigation. Bypasses don't make very good investigations.

Knowledge check 41

Why are playing fields and golf courses regarded as good use of the rural–urban fringe?

2.6 How are rural settlements changing?

How and why do rural environments vary in character?

There are an enormous variety of rural environments and settlements within them. In most MEDCs people are **psychologically urbanised** because they expect the same services in rural areas, e.g. broadband, cable television etc. The factors affecting the nature of a rural settlement can be the product of: distance from an urban area; strength of planning controls; location in the country; whether the area is conserved by legislation, e.g. in a National Park; land holdings in the area; the architecture of the settlement.

The nature of rural settlements

A number of terms are used to describe the nature of the settlements that form an important part of the rural landscape. You will be familiar with most of these.

- **Distribution** — how farms, villages and dwellings are spread across the landscape.
- **Density** — the number of people, farms, dwellings and settlements per km^2.
- **Hierarchy** — a functional system based on the importance of each settlement in an area. Importance is a function of population size and range of services. The higher up the hierarchy, the lower the number of settlements and the further apart they are.
- **Dispersed** — settlement is in the form of scattered, isolated farms and dwellings.
- **Nucleated** — dwellings are concentrated together to form hamlets and villages in favoured sites, for example near a spring or at the junction of mountain valleys. The shape of the nucleated village can be linear or compact.
- **Site** — the nature of the land on which the settlement is built. Favourable sites are often flood free, south facing, sheltered, on a spring line, etc.

- **Situation** — the location of a settlement in relation to its surrounding area.
- **Function** — the purpose of a village, e.g. agricultural, leisure, mining, fishing and tourism.

There are certain characteristics that are common to most rural areas. It is possible to develop an **index of rurality (ruralness)** based on whether an area shows these characteristics or not. In 1971, Cloke identified 10 key indicators of rural life for MEDCs, based on census data. Table 9 lists the indicators that apply to rural settlements and assesses how useful Cloke's indicators are today.

Table 9 Cloke's Index 1971 (revised)

Revised 1971 indices	Characteristic 2011	Change since Cloke wrote index in 1971
% of workers in agriculture and forestry	Very low in villages	Declined since 1971
% of population over 65 years old	More retired in rural areas	Distinguish rural retired from immigrant retired
% of working age population 16–65	Varies with distance from employment centres and proportion in higher education	Counter-urbanisation brought more workers into rural areas
% population change among resident population	Rising due to in-migration but falling in remote rural areas	Second home population has risen — more in remote areas
Distance from urban area	Greater the distance, the more rural	Still true
% employed outside of settlement	High in commuter and counter-urbanised areas and lower further away	Teleworking, working from home one day a week, are altering this
% resident less than 5 years	Measures recent immigrants; may decline if no school easily accessible.	Slight increase but school transport withdrawal may affect families
Population density	Declines with distance although rising with infilling. Lower in remote areas	Still same as 1971
Occupancy rate: % of population at 1.5 persons per room	A general trend not just in rural areas that shows decline	Measures little in 2011. Time distance to service provision would indicate more
Household amenities (baths, central heating)	Better measures would be car ownership and access to mobile phone network	A dated measure in twenty-first century

Cloke's criteria did identify a spectrum of rural settlements but, as the right-hand column shows, many of these criteria are inappropriate 40 years later. Alternative criteria might include:

- Ability to receive signal for mobile phone.
- Access to broadband.
- Access to a supermarket, secondary school, and health care.
- Percentage of built-up land in parish.
- Access to mains gas — one in three rural properties in 2008 did not have mains gas.
- Percentage owning two cars (now a necessity for many rural families).

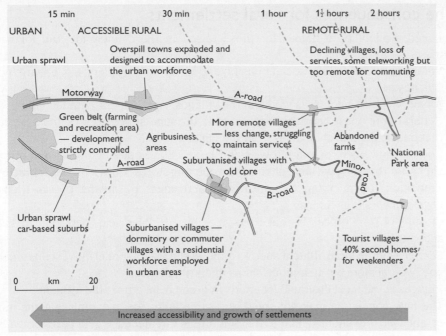

Figure 31 The range of rural settlements with distance from a city

Changes in rural population and settlements

Although, in general, remote rural areas lost population in the period 1930–1970, in most MEDCs, accessible rural areas gained population as a result of the movement of people and employment from urban areas. The movement was facilitated by improved public transport and increased ownership of cars. From 1970, as **counter-urbanisation** began to take place, most rural districts began to experience an upsurge in population known as **rural turnaround**. Rural areas (even the more remote rural areas) increased their population by an average of 16% between 1971 and 1991. This is in contrast to metropolitan cities where populations decreased by 18% in the same period.

Counter-urbanisation

The factors that led to counter-urbanisation vary from settlement to settlement but the following are among the most common explanations:

- The planning process that created new towns, expanded towns and permitted small towns to expand, led to more jobs within easy reach of rural settlements.
- Improved road communications (motorways) and faster rail transport.
- Improvements in communication (e.g. broadband and teleworking).
- Growth of branch factories (branch plants) in the new planned settlements.
- People will take jobs in scenic areas, sometimes for lower salaries and will trade off salary for a better environment.
- Green belt legislation that made areas more desirable.
- Diseconomies of scale in cities.
- National shift towards service employment and specialised rural manufacturing growth.

Knowledge check 42

What is the meaning of rurality?

The consequences for rural settlements

Reduced availability of homes for those raised in the countryside and for rural workers.

- Villages planned to take social housing and lower-paid workers, e.g. Lavant, near Chichester.
- A new service-sector class works in the villages, e.g. cleaners, small builders.
- General stores close because people shop in the superstores in cities and have cars that enable them to shop away from the village.
- Counter movement of some towns and large villages attempting to retain local shops, and banning the large chains from opening stores (known as **Tescoisation**).
- Transition towns trying to improve attractiveness by means of eco-friendly developments (e.g. Totnes).
- Closure of post offices because business has declined (e.g. pensions to bank accounts).
- Services for the more affluent such as antique shops — Arundel in West Sussex has one general store but 10 antique shops and art galleries.
- Villages in the most desirable locations in green belts attract the affluent.
- The new population may not speak the same language, meaning that the more established inhabitants feel that their culture is threatened, e.g. in parts of Wales.

Table 10 and Figure 32 summarise the possible effects of change on villages.

Table 10 Why have rural services declined and why might they improve?

	The bad news	The good news
Food shops	• Supermarkets opening even in quite small towns, with lower prices, extended hours and free buses from some villages	• Relief from uniform business rates • Development of new types of village shops, e.g. farm and garage shops
Post offices	• Small offices downgraded to part-time community office status — unattractive as going concerns for people to run • Diversion of pensions business to banks	• New deals for rural post offices with banks, with the post office counters for general new business
Public transport	• Bus deregulation led to closure of uneconomic routes • New unitary councils subsidise their own services, not rural cross-council links • Not enough passengers	• Grants for community taxis and buses which are often developed by parish councils; such subsidies were being threatened in 2011
Village schools	• Ageing population leads to a lack of 'customers' • As schools compete for numbers the tendency is for children not to attend their local village school, or they go 'private'	• Development of nurseries • Generally good performance of small schools encourages numbers • Clustering strategies for a shared headteacher • Government grants for small schools
Libraries	• Costs of provision • Cuts in council services	• Increased mobile libraries
Primary healthcare	• Closure of GP 'branch' surgeries • Decline in NHS dentistry • Essentially escalating NHS costs	• Creation of mini health centres • Extra grants for rural GPs • Development of rural pharmacies • Use of internet for consultations
Village halls	• General trends of changing family habits • Withdrawal of funding for youth clubs and social services	• Millennium grants for refurbishment

Strong retirement element or people who have lived there for years

| 90+ |
| 85–89 |
| 80–84 |
| 75–79 |
| 70–74 |
| 65–69 |
| 60–64 |
| 55–59 |
| 50–54 |
| 45–49 |
| 40–44 |
| 35–39 |
| 30–34 |
| 25–29 |
| 20–24 |
| 15–19 |
| 10–14 |
| 5–9 |
| 0–4 |

Fewer old people

New arrivals who commute to work (91% have car access or work from home)

Waisted shape — young people move away to look for work or cheaper housing

Expanding population — a baby boom from young families who have moved into area

Regressive base — few children born

15%   10%   5%   0%   5%   10%   15%

Declining population structure	Expanding population structure
The village is declining because of out-migration, and the loss of the productive element has led to a further decrease. It is a very ageing pyramid	The village has expanded because of new in-migration which, because of the age profile, is providing further natural increase
Consequences of decline	**Consequences of expansion**
• As villages decline, many of the people left behind are old and of limited means • As people move away because of a lack of access to employment or affordable housing, other services, such as post offices, village schools and buses, become uneconomic, so even more people move away • Houses are bought for second homes, creating a ghost town for much of the year • Deprivation sets in, so many of the people left cannot move away and lead very restricted lives	• Pressure on key services such as school and health • Creation of several new 'small' housing estates, often executive houses local people cannot afford (could be called rural sprawl) • Little impact on village services, as newcomers rarely use village shops or garages — but they often use the pubs • Increased traffic congestion, especially at peak times — many new families have two or even three cars • Many new villagers do contribute to aspects of village life such as parish councils, Women's Institute (WI) and the church, but most do not • Villages are often dormitory villages with little traffic during the daytime • Conflicts can occur between traditional villagers and newcomers — traditional villagers feel their values are not respected
Breaking the spiral of rural decline and deprivation is the key issue	The main problem here is trying to maintain the rural community and a rural quality in the face of increasing urbanisation

Figure 32 Rural decline and expansion — the geographical consequences

Rural poverty and deprivation

Rural poverty is a problem in MEDCs. Rural areas are largely regarded as more affluent than urban areas, but they contain a significant proportion of poorer people, scattered thinly, with small pockets of severe deprivation. Because of the smallness of these groups, and the perception of rural areas as idylls, they can be easily hidden. Rural deprivation is usually associated with a lack of material resources and is really a 'state of non-well-being', including impacts on physical and mental health caused by this lack.

Some peripheral areas, such as rural Cornwall, are now eligible for Category 1 EU development funding (the most extreme hardship) because these areas have some of the lowest average earnings of anywhere in the UK. Equally, while unemployment is generally lower in rural areas, finding employment can be difficult for school leavers. There is often a limited range of job opportunities because of physical constraints, and much employment (around 27%) is part-time and, therefore, lower paid.

Rural areas contain a greater proportion of elderly people (20%) and while housing in rural areas is generally of a higher standard, there is little affordable housing in villages. The poverty of these mainly elderly people can be extreme, especially when it is combined with deprivation of accessible basic services, and the fact that people are less willing to declare their needs. Rural **social exclusion** exists, where people are excluded from social, economic or cultural opportunities because of low income, poverty, poor health or reduced access to services. **Social exclusion** is the process whereby the various systems that should guarantee the social integration of individuals or households fail to do so, possibly due in part to failure of employment provision or failure of housing markets.

Rural deprivation takes three forms:

(1) Household deprivation is concerned with the hardship of individual households trying to maintain a living standard. Examples are the plight of tenant farmers in less favoured areas, such as the Welsh mountains, or that of elderly single people solely dependent on their state pensions. Household deprivation is typified by poor-quality housing and high levels of income support and benefit payments.

(2) Opportunity deprivation is concerned with country dwellers' lack of access to education, health, work, social services and shops. Many rural families have to face the expense and difficulty of travelling longer distances for basic services, such as fuel, as well as paying more for them.

(3) Mobility deprivation is a measure of the lack of transport for going to work or for obtaining basic services, many of which are now concentrated in key settlements because of rural rationalisation of settlements. Owning a car or a motorbike is extremely expensive, especially for low income families. Access to a village taxi, post-bus or community bus service can be vital in the absence of commercial bus services.

The combined impact of the various facets of deprivation can, unless arrested, cause a cycle of **rural decline.** This can lead to rural **depopulation** if the people lost by out-migration are not replaced by newcomers.

Knowledge check 43

What is a transition town?

Examiner tip

Get some examples of these different forms of deprivation especially if you live in a town and have less knowledge of a rural area.

Knowledge check 44

What is opportunity deprivation?

Summary

- The focus of this theme is settlements in the developed world.

- It is essential that you carry examples from your field studies or your reading that enable you to illustrate your answers.

- Settlements are different from one another because they have different economic, social and cultural structures.

- The inner city is a major area of change and is the area where most governmental intervention has taken place.

- CBDs evolve due mainly to market forces causing change. The bigger the city the more functional segregation of land uses there is.

- Beyond the city the rural–urban fringe is a zone of competing land uses and is where there is competition to change the land uses most radically.

- In the fringe and beyond in rural areas settlements are changing as a result of technological, economic and social forces operating on society.

Questions & Answers

Preparing for the unit test

Geographical terms

It is a good idea to collect your own list of key terms (your own geographical dictionary) as you come across them in your studies. Also, build up your own memory bank of simple, effective maps and diagrams. Practise drawing these because, ideally, you need to be able to spend no more than two minutes drawing each one. Finally, you will need your own examples to illustrate the points that you make. You will come across examples and diagrams in this book; remember that you can impress by the use of different, relevant examples rather than the standard ones in your textbooks.

Where to find good examples

Always try to find examples from your own studies in addition to the ones provided by teachers. *Geography Review* and *Geographical* can be good sources of original examples. Websites such as NGfL (**www.ngfl-cymru.org.uk/eng/vtc-home/vtc-aas-home/vtc-as_a-geography**) are useful. Quality newspapers, e.g. *Times, Guardian, Independent* and *Daily Telegraph*, are other sources that provide good examples. But you need to be aware of bias. There is good geography constantly in the media so make use of it.

The unit test

Timing

There are three questions in the exam based on the content of Unit G2. The topics appear in the same order as in the specification. The examination is 1 hour and 30 minutes, and you must answer all three questions. There is, therefore, a mark to be gained every minute. The table below gives you some idea of the timings that you should ideally try to follow. The timings for questions 1 and 2 are identical.

	Activity	Time in minutes
	Read all three questions. Decide on the order of questions to be answered. (The question order below assumes that you answer questions in order.)	1
Q1 and Q2	Study the first resource carefully and answer the 5-mark question Part (a).	5
	Read Part (b) and plan your answer.	2
	Write your answer (probably about one side of prose plus any maps or diagrams).	10
	Read Part (c) and plan your answer.	2
	Write your answer (probably about one side of prose plus any maps or diagrams).	10

	Activity	Time in minutes
Q3	Study the resource(s) for Question 3.	2
	Plan response to part (a).	1
	Write your answer to (a).	6
	Read and plan response to part (b).	1
	Write answer to (b).	8
	Read and plan response to part (c).	1
	Write one-side answer to (c).	9
	Check work.	3
	Total time	**90**

How answers are marked

Markers always assess the *overall* quality of an answer against the marks scheme for that paper. The holistic mark might draw upon qualities from several levels and award a mark that best fits the combination of qualities in the answer. In addition all marks are also based on assessment objectives (AOs) which enable marks to be awarded for knowledge, application of that knowledge, and skills. These may be quality of language skills, or the skills of drawing or interpreting maps, photos and diagrams.

The levels marks scheme used for parts (b) and (c) of questions 1 and 2 and part (c) of question 3

Level	Marks	Descriptor
3	8–10	Very good knowledge and understanding with a good use of examples. Full descriptions and explanations. There is both breadth and depth to an answer that may draw together a variety of points. Follows the commands in the question. Has a good mini-essay style, well written with a good use of the language of geography.
2	4–7	Some knowledge but not always clear that information/principles are fully understood. Information and explanations are partial. Sound descriptions. There should be more detail. Essay style present but probably lacks paragraphing and introduction/conclusion. Some use of the language of geography.
1	0–3	Limited and superficial knowledge with limited understanding. Little use of examples and if they are present it will be e.g. Africa (often quoted as a country when it is a continent). The material will be simple and probably a very brief one paragraph response with no development. Slips of grammar and spelling evident.

Quality of written communication

There are no specifically allocated marks for the quality of your writing. However, you should try to use correct punctuation and grammar, structure answers into a logical sequence with a brief introduction and conclusion, and use appropriate geographical terminology.

Managing questions in a minute

There are guidance notes on the cover of the examination paper. Read these as they stress use of examples, and the necessity for good English. All questions have two major components. There are **command words** telling you what to do with the subject matter, such as outline, discuss, evaluate, assess. Secondly, there is the **subject matter**. Use highlighter pens in different colours to emphasise the commands and the topic.

- **Describe** — show what you know is occurring or what a map or diagram shows. It expects you to know what a process does, what happens, where it occurs, when it occurs. Who or what causes a process and whom it may affect can be part of descriptions.
- **Explain** — say why and how something occurs. It might involve some description but do not rely on description as a means of explanation.
- **With the aid of a diagram** — draw a diagram, which must be labelled. The axes on a graph must also be labelled.
- **Identify** — provide listed points which are supported by examples.
- **Examine** — provide reasons and discuss both sides of an issue: for instance, 'Examine the costs and benefits of regeneration schemes in cities.'
- **Justify** — give reasons.
- **Outline** — state the main points or factors (at least two), with supporting examples.

Examiner's comments

In this Questions and Answers section each student's answer is followed by examiner's comments, indicated by the ⊜ icon. The comments show how the marks have been awarded and highlight specific problems and weaknesses and areas for improvement.

Theme 1 Question 1

Example 1 **Fertility and death rates**

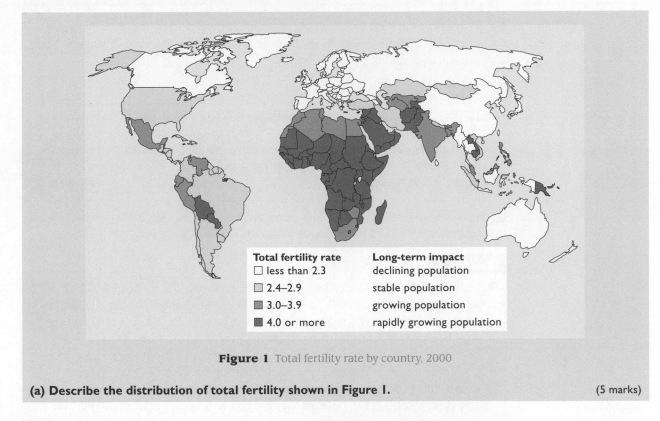

Total fertility rate

☐ less than 2.3

☐ 2.4–2.9

☐ 3.0–3.9

■ 4.0 or more

Long-term impact

declining population

stable population

growing population

rapidly growing population

Figure 1 Total fertility rate by country, 2000

(a) Describe the distribution of total fertility shown in Figure 1. (5 marks)

ⓔ This section tests your ability to interpret a map. Make sure that you can name more than one area.

(b) Give reasons why countries experienced a total fertility rate less than 2.3, the replacement level. (10 marks)

ⓔ The question only refers to the lowest category on the map. There should be more than one reason and there must be valid examples.

(c) With the help of a diagram, describe and explain the changes in the death rate shown by the demographic transition model. (10 marks)

ⓔ This part expects a diagram, description and explanation.

Student A

(a) The African continent has the most rapid growth **a**. Countries which have a total fertility rate of 2.4–2.9 are mostly MEDCs **b**. These countries have a stable population as their standard of living is higher. Declining population is also found in MEDCs.

ⓔ **2/5 marks awarded. a** This is a basic, descriptive answer that only identifies one area. **b** It accounts for one category, which was not asked for in the question. Therefore, it lacks any depth of description and would only gain 2 marks.

Student B

(a) Rapidly growing population can be seen in much of Africa, Saudi Arabia and Pakistan and parts of Latin America. The population is growing less rapidly in India and Central America. The areas of stable population are North America, mid Eurasia. The countries with declining population are in Europe, North America and Australia. **a b**

ⓔ **5/5 marks awarded. a** This answer covers all of the categories and **b** provides examples of each category. It will gain full marks despite the generality of the description.

Student A

(b) Many countries experienced a total fertility rate below the replacement level (less than 2.3) because in those countries the government is placing policies where families are not to have a large number of children. The countries where total fertility rate is below the replacement level of 2.3 are countries which are developed, such as Europe, Australia and some places in South America **c**. The government places different types of policies, for example, in China, there is a 'one child policy' **a**. This was placed because the population was very large and growing every year, but now they are controlling the population. The 'one child' policy is that if a family has not more than two children, they get a 5–10% incentive in their salary, but if a family has more than two children, they get a 10% reduction in their salary **b**. Also in some countries **d** people decide not to have any children because they want to work and have money to spend on themselves. They use contraceptives to stop having children **a**.

ⓔ **5/10 marks awarded. a** The answer provides two reasons. **b** Too much time is spent on the China policy rather than providing more evidence of other policies. **c** It makes the basic mistake of calling Europe a country without naming a country here or in South America. **d** The language does include repetition and only develops the points very generally. It does not refer to replacement level. There are no paragraphs in this mini essay, which is consequently rather short. It is a Level 2 answer and would gain 5 marks for its content and reasoning.

Student B

(b) Most of the countries experiencing a total fertility rate of less than 2.3 are the
MEDCs **b**. Reasons for these countries having a total fertility rate below the
replacement level are because there are higher percentages of women working
in the work force in either full time or part time jobs **a**. With their involvement
in the workforce, women tend to want fewer children and at a later age as
they want to concentrate on their jobs **b**. In MEDCs, the healthcare system and
pension schemes are very good, therefore decreasing the number of births as
most children will make it to adulthood and parents would not need children
to look after them when they are elderly as they will have a pension scheme
a b. Contraception and birth control measures are readily available, therefore
reducing the amount of unwanted pregnancies **a**. Family planning schemes are
also available, therefore enabling families to plan the number of children they
want to have **a c**. Many people stay on in further and higher education, which will
lead to them marrying at a later age and having fewer children as they want to
concentrate on their careers first **a b**.

Sometimes there are government policies to stop or encourage people to have
smaller families.

ⓔ **8/10 marks awarded.** This mini essay is the right length for 10 marks taking about one side
of an answer book. **a** It covers a range of factors which are well explained in a theoretical sense.
b It would benefit from better paragraphing and some words of introduction and conclusion.
c It is difficult to provide examples of these factors, although naming countries where this is the
case would help. Other factors that could be used are: societal norms for size of family, lifestyle
preferences, costs of children, delayed parenthood and increased infertility, and singletons. It is a
Level 3 response and would gain at least 8 marks and be on the way to an A grade.

Student A

(c) *Demographic transition model*

The DTM shows that in stage 2 the death rate drops causing the population to rise **c**. The DTM shows the process countries go through in terms of economic development (the different stages) and its effects on population. As countries develop they can afford better technology and food. This means there is better hygiene and healthcare, so death rates drop until most deaths are caused by old age or accidents. As the DTM shows, this affects the population bringing death rates below birth rates and causing population growth. Another factor causing the low death rates in MEDCs which are in stage 4 of the DTM is because of improved medical facilities, diet is balanced and due to that water and food is not contaminated **c**. Hence diseases such as cholera and typhoid can be avoided. In stage 3 there is a steady death rate **c d**. This means that the country will have a steady population growth. This is common in MEDCs such as the USA **e**.

ⓔ **5/10 marks awarded.** The student has not assembled the response into a coherent answer. **a** The diagram is not accurate, especially for Stage 3. **b** There is no mention of Stages 1 or 5 despite them being on the diagram. **c** Stages 2–4 are dealt with in a random order but the reasoning is on the right lines for Stage 2 . **d** Stage 3 text is partly correct but the example **e** is incorrect unless it is given a date such as 'late nineteenth century USA'. The answer lacks a full sequential description of the death rate line but does get credit for some of the explanation. It is a Level 2 response gaining 5 marks.

Overall, this student will gain 12/25 marks, which is very much in the middle of the marks range and indicative of a C.

Student B

(c)

a

In the demographic transition model that I have drawn it shows that death rate has changed over time and many countries fit into the graph in each of the stages **b**.

In the graph the death rate fluctuates very highly with the birth rate in Stage 1. This tends to occur during a war like during the Napoleonic wars and earlier. Countries in Africa such as Rwanda and Congo fit into stage 1 because of their high death rates caused by war, AIDS and starvation and a lack of access to clean water **c**. In Stage 2 the death rate drops. This is from an improvement in medical care which saves lives and stops deaths **c**. In stage 3 the drop in death rate slows down but it is still slowly dropping, this is because of the last of the older generations from when the death rate rapidly declined in stage 2, but the birth rate remains high **c**, **d**. This was when people started to know how to stop disease and help keep people alive so that they lived longer **c**. Stage 4 shows the death rate staying constant with a bit of a rise, this is caused by a famine or disease like when flu hit the population at the end of the first world war **c**. Stage 5 the death rate rises above the birth rate to show decreasing population, this is just a guess as stage 5 has not properly occurred yet. Many countries fit into different stages like now Australia, UK, Sweeden **d** etc. Italy and Japan also fit into stage 5 because they have ageing populations

ⓔ **8/10 marks awarded.** **a** The diagram is fine. **b** The student attempted a brief introduction but no conclusion that summed up the various explanations. The description is reasonably accurate. Explanations are present although the quality of expression can give the marker the idea that he/she doesn't understand. **c** The examples are all reasonably accurate. **d** The student's quality of English is variable and the fact that Sweden is misspelt at the end will be ignored by the marker although it will influence the quality-of-language aspect of the marks. There is enough description plus the diagram, which, when combined with the explanations, will just get this student a grade A despite the slip in spelling Sweden. It is a Level 3 response worth 8 marks.

Overall, this student gains 21/25 marks for the question, which is an A grade.

Example 2 Population growth, migration and mortality rates

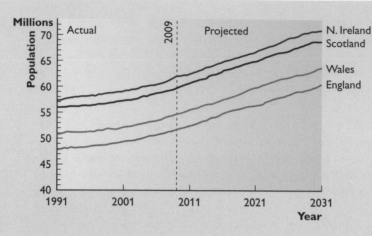

Figure 2 UK population 1991–2031 (source: www.statistics.gov.uk)

(a) Describe the patterns of population growth shown in Figure 2. (5 marks)

ⓔ This question requires you to demonstrate that you understand a simple graph and can interpret the data.

(b) Outline the consequences of the inflow of refugees and asylum seekers into countries with developed economies. (10 marks)

ⓔ 'Outline' requires you to try and give as many consequences as possible. You will need to define the terms and provide examples from more than one country.

(c) Explain reasons for variations in mortality rates between countries at **TWO** different stages in the demographic transition. (10 marks)

ⓔ The question expects explanations for variations. Ideally you should have data from one country at each stage that you select. Examples drawn from more than one country will be accepted.

Student A

(a) Figure 2 shows that the population is growing and is projected to grow in England, Wales, Scotland and Northern Ireland. England's population has risen from nearly 48 million in 1991 to around 52 million in 2009 **a**. It is projected to rise to over 60 million in 2031. The total population of the UK has risen from over 57 million to around 60 million in 2009. It is projected to rise to 71 million by 2031 **a**.

ⓔ **2/5 marks awarded. a** There is only a description of two of the most obvious aspects of the diagram. The fact that it is a cumulative total is not noted and no comment is made about the other totals. Therefore, this response will only gain 2 marks.

Student B

(a) Figure 2 shows patterns of population growth from 1991–2031 **a**. It shows that each of the countries in the UK has experienced a rise in population from 1991 to the present and all are expected to rise even further by 2031 **b**. In England the population in 1991 was 48 million **b**. This has risen now to 51 million and is expected to rise further to 61 million by 2031 **b**. Overall in the UK the population was 57 million in 1991 **b**. It has risen to 62 million now but by 2031, is predicted to reach 71 million people **c**. It is also interesting to note how the growth rate will increase in this time also **c**. In the ten years from 1991 to 2001, the overall population of the UK rose by 2 million, however in the ten years from 2021 to 2031 the population is projected to increase by 5 million people **b**.

ⓔ **5/5 marks awarded. a** The answer shows that the student understands the diagram. **b** A range of valid points are made about absolute and English totals. **c** In addition, the rise in the numbers was noted and described with support. In the time allowed this is a very good response. This answer will gain 5 marks.

Student A

(b) In Caia Park in Wrexham **a**, although the majority of people are from a white-Welsh background there is a small number of Iraqi Kurd refugees living in part of the estate. One Iraqi Kurd was subject to an unprovoked attack by 15 men with iron bars and knives **b**. The next evening there was a riot which started in the Red Dragon Pub in which more than 200 people were involved **b**. Petrol bombs were thrown and cars were set alight **b**. These events highlight the conflict that is often created when refugees and asylum seekers **c** move to a country such as the UK with a developed economy **e**.

Other problems created are languages. Services such as schools and doctors may have to pay more money for translators to cope with the refugees and asylum seekers. This money often comes from taxes and could cause tax rates to increase. **d e**

Refugees and asylum seekers may also not be able to get any work in countries such as the UK because of the language barrier and many of the refugees and asylum seekers' qualifications may not be accepted. This gives them a big disadvantage when looking for a job. **d f**

ⓔ **4/10 marks awarded. a** The example is good but it is about the flow into one area and **b** the consequences are implied to be rioting. The example could be used more constructively to make a point. **c** There is no attempt to define the terms. **b** We are not told what the conflict is about and it implies that all consequences are negative. **d** There is no evidence provided in the last two paragraphs although the points may be valid. There is no reference to any gains for a country. **e** The student gives the impression that he/she has not read much geographical literature on the topic and is relying on the media with all of its biases. The answer is a mixture of vagueness and detail. **f** There is no conclusion to the essay, which will gain 4 marks.

(b) There are many consequences due to the inflow of refugees and asylum seekers **a** into countries with developed economies. In the UK, a total of 800,000 Eastern Europeans have entered the UK, with 500,000 of these from Poland **b**. Although many of these are economic migrants, there are some refugees and asylum seekers from the former Yugoslavia in the figures **b**. They bring with them many consequences: quite often, they don't speak the language of the country they are moving to. This has been a problem at Beeches Primary School in Peterborough, where there are 700 students, 25 different nationalities. Many translators are needed, costing the school and indeed the government money **c**. Also, in some areas of the UK where there are not enough jobs, they can be seen as taking people's jobs **c**. Quite often, though, they are the jobs that local people don't want to do such as flower picking.

They are cheap to employ, and are willing to work longer hours than local people. A further positive consequence is cultural benefits they may bring with them **c**. These could be in the form of different religions, to fast food takeaways **c**. Finally, the inflow may cause racial tension in many areas, and could form cultural suburbs and ghettos which would be bad for the local community **c**. Another example of an inflow of refugees and asylum seekers into developed economies would be West Africans that travel through Libya (intervening place) to reach Europe where similar consequences are encountered **d**.

ℯ **7/10 marks awarded. a** There are no definitions. **b** The examples could be made more relevant if the answer focused on Croats and Serbs rather than Poles. **c** The consequences have the correct principles but not the best support in terms of the groups that are discussed. **d** The final sentence does suggest that the student knows what a refugee might be although early on the answer is more about issues of immigration in general. Nevertheless, the principles will enable this answer to gain 7 marks.

(c) In Italy, mortality rates are very low because there is a lot better healthcare than in a country such as Tanzania, which is at a lower stage of the demographic transition model **a**. A free public health service was set up in 1978 in Italy. This allowed people who up until then could not afford good healthcare to get it. This increased life expectancy and lowered mortality rates **b**.

Another reason that mortality rates are low is that since 1945 Italy has had a lot of industrialisation and urbanisation **b**. This means that the economy of Italy has grown considerably and therefore more people are richer. Generally **d** this means that the mortality rate is lower as people are able to afford a high standard of living **b**.

In Tanzania there have been many problems with water-borne diseases such as cholera because of the lack of clean water in many areas of the country. This has increased the mortality rate significantly and has been one factor why the life expectancy is just 47 years **b c**.

Tanzania is generally a much poorer country than Italy. Therefore, many people are too far away from medical facilities and cannot afford to travel there or pay for the treatments. This leads to many people dying when if they had lived in Italy they may have survived.

ⓔ **5/10 marks awarded. a** The introduction names two countries but does not state the stage each country is in. **b** Two reasons for low mortality are stated but the student doesn't say which stage of the model the answer is referring to. Do not assume that the examiner knows. Tanzania would be a good contrast if we knew the stage of the transition that was being discussed. **c** There is only one figure quoted in the whole answer, which should have contained figures illustrating the different death rates. **d** The answer is too general (even the student used the word 'generally') although the explanations have some validity. This response is awarded 5 marks for the generalisations.

Overall, this student would probably score 11/25 marks, which would be at the C/D border. Precise data, a knowledge of the DTM and better explanations would raise this mark.

Student B

(c) Namibia and the UK are at 2 different stages of the demographic transition model. Namibia is at stage 2 of the model meaning that their mortality rate is very high, 20/000 **a b**. This is due to a number of reasons. Namibia is an LEDC, meaning that their economy isn't as developed as those such as the UK **a**. Because of this, there isn't as much healthcare, with medicines not always readily available, and this affects the mortality rate. Also education isn't as successful as in MEDCs, so people aren't always educated or made aware of sex and disease **c**. The lack of contraceptives means that AIDS is a common cause of death in such countries, also affecting the mortality rate **c**. In the UK which is at stage 4/5 of the Transition Model, healthcare is readily available and education about diseases such as AIDS is widely available, not only through schooling but through advertising such as on TV and the internet **c**. Also food is more readily available whereas in Namibia food isn't as readily available and many people rely on the help of Aid Agencies and donations. Because people live longer in the UK (life expectancy 79 whereas in Namibia it is 50 the UK) is definitely in Stage 4 where the death rate is 6/000 **b**.

ⓔ 7/10 marks awarded. **a** This answer does provide some explanations for the high death rate in Stage 2 as opposed to Stage 4. **b** It quotes some figures in support, which are of the right order even if they are not entirely accurate. The material on the UK is lacking, perhaps because time was running out. **c** Nevertheless, the principles are explained briefly and in relation to the model. The response merits 7 marks.

Overall, with 19/25 marks, this question should place the student above the A/B border.

Theme 2 Question 2

Example 3 Social structure and land uses

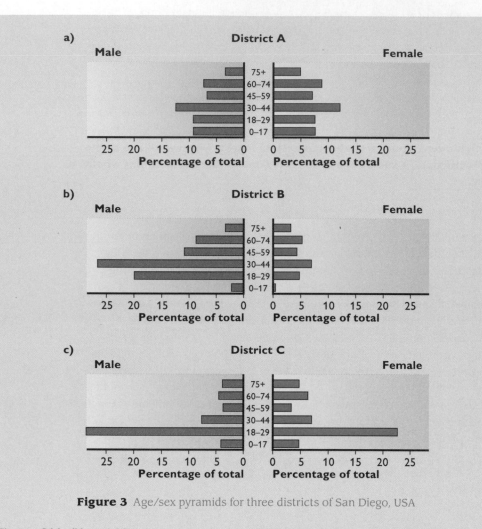

Figure 3 Age/sex pyramids for three districts of San Diego, USA

(a) Study Figures 3(a), (b) and (c), which show the population pyramids for three districts of San Diego, USA. Suggest and justify the possible locations within the city for each of the pyramids.

(5 marks)

ⓔ This demands an understanding of population pyramids (Theme 1) and social structure of cities (Theme 2). 'Justify' requires you say why you have made a suggestion by providing brief evidence.

(b) Outline the economic factors which help to create different residential districts within a large urban area.

(10 marks)

ⓔ This requires you to give as many reasons as possible. Although it says a large urban area, you should try to provide examples from one or more cities.

> **(c) Explain how and why redevelopment is taking place in the CBD.**   (10 marks)

ⓔ 'How' needs you to refer to actual examples and 'why' asks you to explain the changes. The focus must be redevelopment.

Student A

(a) District A is the inner city whereas district B is an area of immigration of males **a b**. There are many young people in District C which may be student housing **a b**.

ⓔ **3/5 marks awarded. a b** The response is correct but brief and is awarded only 3 marks because it justifies without evidence.

Student B

(a) District A appears to be an established suburb because it has a broad age range of population **a**. District B has few females and could be anywhere where there are mainly males moving in **b**. This could be an immigrant area or an area for the Gay population **b d**. District C is dominated by 18–29-year-olds which suggests it might be an area where the young first live after leaving home **c d**. They could be students.

ⓔ **5/5 marks awarded. a** This answer about District A is basically correct and justified as it is a residential area in the suburbs. **b** District B information is almost correct but it is not locating the pyramid although the reasons are hinted at. It is close to the CBD and is an area housing young upwardly mobile males — not nest-builders as there are few children. **c** District C is correct; it is an area around the university where studentification has taken place. **d** There are several justifications in the answer.

Student A

(b) The price of land is a factor as we see that most cheaper land is found further away from the CBD but houses might get more expensive with distance as rich people don't want to stay near the noise and other types of interference associated with the CBD **a**.

The affordability of transport also plays an important role because if one cannot afford a car or pay bus fares he is going to locate nearer the CBD. The opposite is true because distance will not be a factor so people can locate further away from the city if they like **b e**

This will see high density residential areas being formed much closer to the city centre in order to cut transport costs. The more wealthy people will locate in medium and low density areas that are found much further away from the city **c e**.

The other factor is rents of houses, you find that bigger houses are not nearer to the CBD because of the cost of land. So rents are higher out there than nearer the CBD. This creates different characteristics as flats are more in town than on the outside of town **a d e**.

ⓔ **5/10 marks awarded. a** In the first and last paragraphs this essay does emphasise the role that land costs play (the term bid rent curve could have been used here). **b** The second paragraph refers to costs of transport. **c** The third paragraph could have been a conclusion. **d** The essay only sees the city as inner versus outer and does not see any detailed division of districts resulting from migration, age and ability to pay (students), and even the role of gatekeepers (estate agents who promote areas to gain higher fees). **e** The answer is Level 2 because information and explanations are partial. There are no examples and the language is rather simplistic.

Student B

(b) There are several economic factors that lead to different residential areas in the city. The bid rent curve says that rich people are willing to pay more for space and therefore the outer city tends to attract people with more money **a**. In contrast poorer people can afford less space and are concentrated in terraced houses as in inner Swansea **b**.

Proximity to good transport links will cause the price of property to rise and so those areas with good public transport or good, fast road links to the city centre will only be affordable by people with money **a**.

The ability to get a loan for a mortgage will also lead to the areas of larger houses only attracting the middle classes. House agents can make places more middle class by only sending the middle classes to look at properties there **a**.

Proximity to good schools will raise the price of property and so only those who can afford the houses will be able to buy near a school. This again segregates people **a**. This happens in many towns such as Kingston **b**.

The richest people often want to live very close to the city centre in regenerated areas of old warehouses or new apartment blocks such as the area around Cardiff Bay and London Docklands **b**.

ⓔ **7/10 marks awarded. a** This answer has a range of economic factors some of which **b** have examples added. The response could have a few more examples or improved detail to gain an even higher mark. Nevertheless, it is an A-grade response, for its understanding of the factors.

Student A

(c) Redevelopment is taking place in the CBD for many reasons. First, the CBD is old and people want new buildings for shops and offices **a**. Old shops are being knocked down and new ones built like Primark **b** because people want cheap clothes because they cannot afford more **c**. Sometimes you get big department stores that are new such as the John Lewis one in Cardiff **a b**. They had to knock down buildings first. City centres have lost shops to out of town shopping centres. Some shops have closed due to the credit crunch **a**.

People like walking around shops and so the council has pedestrianised the main shopping street and sometimes they have covered the road to make it more pleasant to shop there **d**.

There are now many offices in the CBD in skyscrapers. They are big because the companies that use them are big and need a lot of space **e**. People journey to work from the suburbs and it is easier to reach the town centre. Nightclubs are always in the CBD because young people want nightclubs **f**. There are many changes in the CBD **g**.

ⓔ **5/10 marks awarded. a** This is a poorly composed answer yet it does demonstrate some knowledge. **b** It does suggest that new shops are coming into the CBD and gives store names. **c** It notes competition for the CBD but does not explain it. **d** It has a section on pedestrianisation and **e** some sentences on offices with some reasoning. **f** Recreation is mentioned but without any reasoning in any depth. Only one place is mentioned although the whole answer is probably referring to Cardiff. **g** The concluding sentence hardly pulls the answer together. It describes the changes rather than saying why the changes have taken place. The four changes mentioned would enable this response to gain 5 marks. There is not enough explanation to gain more credit.

Student B

(c) Redevelopment has taken place within the CBDs of many urban cities **a**, including Birmingham in the last 20 years as a way of bringing back shoppers and residents. In the 1970s and 1980s Birmingham went through deindustrialisation and this led to large unemployment and derelict factories which became retail sites. The communities such as Aston also went into deprivation **d**. This scares in the inner-city part investors off from the city and they chose to build new houses and shopping facilities as well as other services on the cheaper more accessible and aesthetically pleasing greenfield sites **b**. This also leads to the decentralisation of retailing where large out-of-town shopping facilities were built away from the CBD. In the Birmingham conurbation, Merry Hill in Dudley, Marshall Lake in Shirley and Fortin Stourbridge are all examples of decentralisation of retailing **c**. The Birmingham CBD was redeveloped with the flagship being the new Bull Ring shopping centre, employing over 8,000 people and costing over £500 million, the mail office in the zone of assimilation as well as the high quality apartments to attract residents to the CBD **a b**. The redevelopment and gentrification **a** of Birmingham's CBD was an attempt to bring shoppers back to Birmingham, the Bull ring has 370 million visitors a year **c** and invites wealthy people to live there. This will hopefully boost the status of the area **b**, further encourage development and stop inner-city deprivation and decline.

ⓔ **7/10 marks awarded.** This essay has identified **a** how and **b** why a selection of changes has taken place and **c** provided some detail regarding retailing. It also refers to aspects of residence in the CBD. **d** Some information drifts away from the CBD. The knowledge shown is worthy of an A grade, although the quality of language and lack of paragraphs does detract from the answer. The detail does enable the answer to gain 7 marks.

Example 4 Perceptions of rurality

Richard Rowbottom

Figure 4(a)   Figure 4(b)

(a) Use Figures 4(a) and (b) to suggest why perceptions of rural life are powerful pull factors in counter-urbanisation.   (5 marks)

> ℮ This question is testing ability to interpret photographs. Marks will be gained if you refer to the photos when answering the question.

(b) Outline how perceptions of rurality have led to changes in the nature of rural settlements.   (10 marks)

> ℮ This a question where supporting evidence is essential for the highest marks.

(c) Explain why the rural–urban fringe is changing.   (10 marks)

> ℮ A definition of the rural–urban fringe would be useful. You need to outline some changes and say why they are taking place.

Student A

(a) Counter-urbanisation is when people move out of the city and into the countryside **a**. As shown in Fig. 4(a), rural villages are seen as clean and tranquil places **b**. Pavements have no litter and the vegetation is neat. Traffic levels are perceived to be low. People perceive rural life to include cottage-like houses with large gardens **c**, a drive and maybe a garage as shown in Figure 4(b) **b**.

People perceive rural areas to be more civilised and suburban lifestyle includes having a local pub and familiar neighbours (Fig. 4(a) **b**. As houses may have private drives (4b) people can park safely and easily in rural settlements and still be able to travel to work.

Figure 4(b) shows a pretty garden and 4(a) a pleasant clean community **b** and people expect all rural lifestyles to be like this and it is an increased ideality **c**.

ⓔ **5/5 marks awarded. a** This begins with a definition of counter-urbanisation. **b** It is a long response for 5 marks which uses the resources to make points as instructed in the question. **c** It even recognises perception, despite the use of 'ideality', the meaning of which is not clear.

Student B

(a) The photos show a typical village with a quiet street running through it **a b**. There are many old buildings with many chimneys showing that they are old **a**. People like moving to areas like this where there is a pub **d**. This village will attract people with a lot of money, which is what counter-urbanisation is, rich people going to villages **a c**. They can commute by car to the big towns. **d**

ⓔ **3/5 marks awarded. a** The points made are generally along the right lines but they are never expressed that clearly. **b** There is only one direct reference to the figure. **c** The definition of counter-urbanisation is incomplete. **d** The answer is too descriptive and lacking enough linkage to counter-urbanisation and perception to gain more than 3 marks.

Student A

(b) The perceptions of rurality have changed for different places. The case study we studied was Framlingham, **a** which had a dramatic population increase. From 1981 to 2001, the population increased as much as 42% from 2,190 to 3,114 and there was an estimate that it would increase up to 3,320 by 2006. House prices increased as much as three times in 17 years from £60,000 for a three bedroom semi-detached house, now the same house is worth £180,000 **a**. People perceive places like Framlingham to be the 'chocolate box' place where everything is perfect and top form but, in reality, the place may not be like that **b**. There are problems associated with 35 hectares of open land lost from expansion of over 646 new houses. Also, traffic flows have increased to and from the village **c**. From the current number of houses being created there has been an increased amount of flooding and overland flow from the new houses being built on greenfield sites and stopping the natural drainage from working **c**. People perceive places to have nice weather but after everyone moves in their views change. Young people have been forced to move out as housing can be expensive and up to 10 times their average income.

ⓔ **8/10 marks awarded. a** This response outlines the changes in the nature of the settlements quite well by using an example reasonably effectively. **b** It lacks more information on perceptions of rurality other than the one reference to chocolate box villages. People perceive quiet, community, basic services nearby, and safety, as well as increasing value of property. **c** It uses a good example and it does outline the changes, one of which is environmental, thus linking changing physical environments to changing rural settlements.

(b) Many people have the perception that rural areas such as Shropshire have relatively low house prices **a**. However, due to so much demand for these rural houses the average house price in Shropshire has risen greatly in recent years **a**.

Furthermore, many people also move their families out to a rural area **a** with the perception that the area will be very quiet with very little traffic. However, due to more people owning 4×4s it is now a lot more busy in these areas **b**. This is backed up by the fact that 55% of road related deaths now occur on rural roads. Therefore, the quiet and safe perception of rurality has now changed for the worse **a**.

In addition, the perception of having lots of time and space to enjoy the countryside may also have changed due to many people living there having to commute miles to work in an urban area every day. This can also leave wives or children alone for long periods of time in a very isolated location **b**. Therefore, the life with a strong sense of local community **a** can also often be proved incorrect **a**.

ⓔ **5/10 marks awarded. a** This answer refers to perceptions although it tends to outline these in general without supporting evidence from a real place. Just saying Shropshire does not illustrate a rural area *per se*. It would be better to look at a specific place such as Shawbury and use this village to illustrate the points made. **b** Some of the ideas are vague, e.g. the points about traffic and even the implication that women and children are left on their own.

(c) The rural–urban fringe is the area of land between the urban areas of the CBD and the rural areas **a**.

The rural–urban fringe (RUF) is changing because it is a greenfield site because it used to have villages within it. A greenfield site is better for developers because nothing has been on the land before so nothing could have affected the land to be built on **e**.

The growing population is causing the RUF to change because more houses are needed for the population **b**. More areas for the families to use such as parks would also be why the RUF is changing from rural to urban **e**.

The land is cheap to buy for developers to build on it so this is the advantage of the RUF **d**.

As the population is growing **b** there are more facilities that are needed such as schools, colleges, hospitals shopping and leisure **e**. This is like Truro's RUF which has all these facilities and housing some people who work locally and have their amenities in the same area **f**.

The RUF is also changing because families are moving out of the CBD and inner city areas so want a place to relax and for the children to grow up **c**.

A few years ago the RUF would have been a place of villages but now as the population grows **b** people want more facilities and a better quality of life so the RUF has changed to provide this.

ⓔ **6/10 marks awarded. a** This answer attempts a definition that is partly correct. It attributes change mainly to **b** population increase and **c** migration. **d** Housing demands are noted, as are **e** other uses. The pressure from developers is introduced early on. **f** There is one example that lacks detailed support. The examiner will mark on the reasons given, which are generalised, and ignore the slips such as migration from the CBD and the use of CBD in the definition. This will gain 6 marks, which is just below the grade A boundary.

Overall, this student will score 19/25, which is just within an A grade.

Student B

(c) Rural settlements are changing **a** due to the fact that there is a change in there **b** population as there are less farmers and more richer people moving in **c**. This is, as there is more popular demand for holiday homes or weekend homes where people who work in the city come and live in the countryside at the weekend **c** **d**. There are benefits and down sides to this change in rural settlements **a**, the benefits are that as more people are moving to these smaller rural towns there is a need for better road quality and accessibility which inturn **b** is better for the locals and the commuters. Anouther **b** upgrade is the fact that as the population increases there will be more shops which inturn **b** will boost the local economy. There are large negatives to these changes **a**, as roads get better people will use the rural settlements as short cuts to get to the larger towns. The biggest issues are with this **b** are the influx of residents and house prices will rise to a point where local wouldn't be able to afford to buy in there **b** own village in the fringe **a** **c** and secondly with new businesses coming into the villages and driving the old ones out **d**.

ⓔ **4/10 marks awarded. a** This answer is very generalised and only refers to the rural–urban fringe in passing because it assumes that changes to settlements can be equated with the fringe. **b** Grammar and spelling are poor. **c** There are some valid points made in a basic fashion. **d** Some points are without evidence. The answer will gain 4/10 for the generalisations, which is a D grade at best.

Overall this student will be awarded 12/25, which is a low C grade.

Theme 3 Question 3: Research including fieldwork

Example 5 Investigation into inner-city change

100 metres

N

Figure 5 Vertical air photograph of part of an inner city in the UK

Study Figure 5, which was used in an investigation into inner-city change.

(a) Describe the variety of land uses seen on the photograph. (7 marks)

ⓔ This question tests your photo interpretation skills. Make sure that you locate the land uses and their variety.

WJEC AS Geography

(b) Suggest and justify how you would plan a questionnaire survey to investigate ONE of the following aspects of inner-city change: (i) environmental, (ii) economic, (iii) social.   (8 marks)

ⓔ You only have to study *one* aspect — if you write about others the examiner will only take the best response. There are two elements: suggesting the questionnaire and planning how to use it.

(c) Evaluate an investigation that you have undertaken into a changing human environment. You should clearly state the question that you have investigated.   (10 marks)

ⓔ The question is asking you to assess the whole project — where it worked, where the results were reliable and where it fell down. Do not forget to name the study.

Student A

(a) Towards the NE **b** there is a small park area **a** for residents to walk through and use for leisure. Just to the south **b** of this area is an area of allotments. In the south there are a large number of parked cars possibly 150 m from a station **a**. However, they do seem to be near large buildings which are possibly used for retailing or business. To the SE **b** there appears to be an old factory which once had rail access by the outline of its site. Just above **c** it there is a newer factory **a** with lorries parked. Much of the photo is covered by residential areas with an area of terraced nineteenth century housing **a** for workers in the factories. Elsewhere the housing is semi-detached **a**. There is a church on the west side **b**.

ⓔ **7/7 marks awarded. a** The answer provides enough detail on a series of land uses to gain the full 7 marks. **b** It uses compass points rather than top and bottom **c** although it does drift into 'above'.

Student B

(a) The majority of land uses in this photo seems to be housing as the centre of the photo and NW are housing **a**. There is a mixture of housing in some kind of estate **a**. To the very NE and E a green park **a** stretches down the edge of the photo. In the SE there is a big factory **a** and a road whereas in the SW there seems to be some sort of company **a, b**.

ⓔ **3/7 marks awarded.** The level of description lets this down. **a** Several uses are identified but **b** the language used becomes less geographical as it progresses. This response will get 3 marks for the broad typology of uses.

Student A

(b) A questionnaire survey can only be planned by outlining simple and direct and closed questions piloted before performing the survey. Each of the environmental aspects such as litter, building quality, pollution or air quality should be mentioned in the survey **a**. Possibly the people being surveyed can be asked to rate each of these aspects of change on a linear scale **b**. Just as importantly though the sampling technique will need to be considered when constructing the survey. A random sample **b** could be more suitable for example instead of a stratified or systematic method because this would prevent any biased results **c** and make the questionnaire easier and more convenient to conduct. The number of people that will be surveyed also needs to be decided. I believe 50 persons will provide enough results for analysis **b** but it still will not take too long to conduct.

ⓔ **5/8 marks awarded. a** This is a response about the environmental aspects. **b** It tries to justify the type of sampling, the subject matter and the sample size. **c** It also makes a comment about bias. It does not have to state the question although it might have helped to illustrate good and poor questions. The suggestions and justification made in 10 minutes of writing time would merit 5 marks.

Student B

(b) A questionnaire to investigate economic inner-city change could be planned by firstly deciding what sort of questions you would ask and also who you would ask **a**. You would need to plan because otherwise you might not get a target audience and also you would need to see what questions to ask that are relevant **b**. Also you would need it spread out over an area and not condensed otherwise the results will lack validity **b**. You would need to ask economic questions possibly to restaurants and ask about who they employ and how many and other details like this **c**. Also how much the employees get paid **b**. You might want to find out how much they get paid and if a wage affects living standards as well **b**.

ⓔ 3/8 marks awarded. This is a response dealing with economic aspects. **a** It is an example of a rambling answer which appears to say little but **b** does have some vague ideas about planning. **c** It does state, poorly, that you need to decide the number of questions, who to ask and where to ask. This answer is very weak. Consequently, it would gain 3 marks.

Student A

(c) I investigated how the environmental quality of the Central Business District in XXXXX is affected by pedestrian flows **a**. I conducted a questionnaire, an environmental survey, a traffic count and a pedestrian count for various areas of the CBD **c**. The main limitation of the investigation was down to timing. Considering that each area was only counted for two minutes, the results were not necessarily accurate. Furthermore, as myself and my assistants **b** moved to different areas, we would be measuring the pedestrian counts and traffic counts at different points of the day **c**. Overall there was about a 2 hour delay between the earliest and latest recordings, this would have affected flows. The questionnaire is subject to opinions, it may have been easy to conduct but the data was not completely foolproof **d**. Likewise, counting traffic and pedestrians was an easy method but it is possible that human error would have influenced the recordings **d**. Measuring and recording samples in a stratified approach was very useful **e**. On the other hand I was able to locate pedestrianised and less pedestrianised areas and clearly show how the environmental quality was affected **e**.

ⓔ 7/10 marks awarded. a This answer tells the reader the subject of the investigation. **b** It does note, in passing, that it was an exercise that used fellow students. There is nothing wrong with group work. **c** A range of methods of data collection are described and there is some evaluation of the issues of data collection **d**. The answer does not go on to evaluate the whole exercise although the last sentence implies a degree of success **e**. This response is on the A/B boundary and would gain 7 marks.

Overall, this student would gain a grade A 19/25 because the excellent performance in (a) and the very good performance in (c) would make up for the slightly less strong performance in (b).

Student B

(c) The question that I investigated was 'What factors affect environmental quality in XXXXX?' **a**. We used quite a few methods, which we thought might help us investigate our question on environmental quality. We did pedestrian counts, car counts, questionnaires and bi-polar analysis. The strength of using a questionnaire is that you can get good qualitative data which you can then compare to other data **b**. Also you get lots of opinions which you can compare **b**. A weakness is that people may exaggerate and give socially desirable answers which will affect validity **b**. A strength with pedestrian counts is that you are able to see how populated an area is and how that correlated to people's opinions in the questionnaires. A weakness is that the pedestrian counts can only give you an idea of environmental quality if you have something to correlate with, e.g. pedestrian count and litter count. Another method we used was bi-polar analysis of an area — this is when we had to evaluate an area in terms of ugliness/ prettiness or pollution or overcrowding, usually done on a 5 point scale **c**. A strength is you can get an idea of an area and its qualities, but a weakness is that you are only getting the one person's opinion and that is whoever is filling it **b**. You improve on this by getting more opinions from others. The sampling we used for the questionnaire was when we went into an area and just chose whoever was nearest to us — convenience sampling. This was positive because it was easy to do but a weakness was that it could be subjected to bias from the person asking the questions. Improve this by systematic sampling **d**.

ⓔ 9/10 marks awarded. a The title is stated. **b** This scores more than the other student because it evaluates by looking at the good and bad points of the method used on the same survey as the first student. **c** It also examines an extra technique, bipolar analysis. **d** The last sentence suggests that the student ran out of time. Do use the time guide on pp. 68–69 and don't rush this last 10 mark answer. This answer would receive 9 marks — a very good A.

Overall, the strong response to part (c) has raised the student's final grade for the question to 15/25, a good grade C. Do always bear in mind that a strong answer to a part can help you get a good mark. Do not be put off by the first parts of a question.

Knowledge check answers

1 The challenge comes from rapid increases in numbers, the uneven increase in numbers throughout the world, the resources needed to sustain the population and because migration also contributes to change.

2 2011.

3 Crude birth rate is the ratio of the number of births per annum to the total population. Crude death rate is the ratio of the number of deaths per annum to the total population. Both are expressed per 1000 people.

4 The world is dominated by the population of Asia and particularly India and China. The population in Latin America and Africa is large in contrast with the far smaller numbers in Europe, European Russia, North America and Australasia.

5 The Chinese have a dominant middle aged population with declining numbers of young people. The over-50s are growing in numbers with the highest proportions aged 40–44. More Malays are in the younger groups and more young age groups with an excess of males. The Chinese have the greater percentage over 50 and more age groups with an excess of women. Malays have largest percentage of adults aged 35–39.

6 Infant mortality is a measure of health provision. The highest rates are in countries where disease control is poor and birth control is almost absent. Lower rates reflect maternal heath provision and levels of female emancipation and education. (You could try to offer explanations for the data in all of columns 3–7.)

7 The number in each age group under 40 years old is declining so that 0–4s are half the number of 40–44 year olds. There is a large number of old females over 80. Women are dominant from 70 upwards — partly the effect of deaths in WWII.

8 African numbers are rising fastest whereas the European share is declining and is a falling share of world growth. The Asian share is relatively static although numbers are still increasing. (You should also be able to offer explanations for the changes.)

9 Here are some examples for you to fit into the cells: quantity surveyors to Dubai, Greek Cypriots from northern Cyprus to the south, south Sudanese from Sudan, Filipino maids to Hong Kong, Zimbabweans to the UK.

10 Ravenstein wrote over 100 years ago. Some migrations do still conform to his laws, e.g. people may move locally to find a larger house, and adults do dominate. The points about rural people and industrialisation might be true in LEDCs but not everywhere.

11 This is a case where your personal experience or that of your family could be used, such as the decision on where to go to university. Some factors are important in some cases but not all.

12 Commonwealth links and language are a common theme for several countries. Intra-EU migration could explain others. Political fears may have induced others to move.

13 For example, South Koreans have congregated in Merton and Sutton, Greater London. Often people migrate to where they know there will be support of family, culture, religion, schools and shops. There is a sense of belonging. French in Kensington and Chelsea, Turks in Enfield, Bangladeshis in Tower Hamlets are other examples.

14 The spatial separation of different groups of people into distinct areas.

15 For example, Natures Way salad processing factory near Chichester employs mainly East Europeans on the factory floor. It supplies major supermarkets with ready prepared salads and fruit.

16 A refugee is a person unable or unwilling to return to his/her homeland for fear of persecution based on reasons of race, religion, ethnicity, membership of a particular social group, or political opinion.
An asylum seeker is a refugee making a formal application to reside in a country when they arrive in that country.

17 A stateless person is anyone who is not considered as a national by any state through its nationality legislation or constitution.

18 Retailing, administration, health and education — all in the service sector.

19 There is less inequality in the most developed countries and more in those of medium and low development. Exceptions to the rule are Saudi Arabia and India. Female literacy rates are lower in the countries with low human development, which will hinder the ability to gain work and to assist in the literacy of children. High maternal mortality implies that women in some countries are still seen as mothers, which will limit their ability to join the labour market.

20 Canal Street in Manchester is associated with the gay community, but as a recreational area rather than where they live.

21 Some service sector jobs have traditionally been seen as female. Some cater more readily for career breaks for maternity. Others suggest that gender bias and glass ceilings may have steered women into these activities.

22 Migration anticipating, or at, retirement. It can be triggered by early retirement, redundancy or genuine retirement.

23 Low-skilled posts assembling products that have been developed elsewhere.

24 108 boys to every 100 girls (globally it is between 102 and 106 — see page 28).

25 Many diseases kill so many people that they have the effect of reducing the population. Death control is the improved management of the disease so that fewer die from it and, therefore, population increases.

26 Village — a settlement in rural surroundings that was once an agricultural settlement. Villages normally contain some services such as a church, an inn and possibly a shop.
Megacity — a key centre of the global economy whose functions affect the whole world. They are sometimes called global hubs.

27 Urbanisation is the process by which the proportion of the population living in an urban area increases. It was associated

with the industrial revolution but is now a phenomenon of the developing world. Urbanism is the process of becoming urban as people adopt urban life styles, change occupations and depend on the city for most of their life. It can be present in rural areas where people have an urban lifestyle and are psychologically urban.

28 The grouping together of activities either in a region (e.g Silicon Valley) or parts of a city (e.g. insurance around Lloyds in the City of London or shoe shops on a High Street).

29 A US term for the movement of white population from an area as another racial group moves in.

30 Agglomeration of like individuals in an area where property rental prices can be afforded. Proximity to the university. Property not sought after by other groups.

31 Nodes are focal points in an urban area — often transport interchanges — where there is a concentration of activity. Landmarks are features in the townscape with distinctive characteristics that people remember, such as the Millennium Stadium, Cardiff or the Spinnaker Tower, Portsmouth.

32 The private estates of the rich who have created their own defensive space such as Ruxley Heights, Claygate and New Caledonia Wharf, London Docklands. Gated communities are elite, exclusive neighbourhoods. They can be found in ex-colonial settlements.

33 Deprivation that has several dimensions — social (high incidence of crime), environmental (poor quality housing) and economic deprivation (low household income/dependency on benefits). It can found in small areas of cities (high-rise blocks in the UK) beside individual streets in the inner city or in parts of social housing on the periphery. It can also be found in pockets in rural areas.

34 Building on sites within a built-up area that had been used previously, such as old railway yards, barracks, industrial sites.

35 Promotional programmes designed to change outsiders' perceptions in order to attract inward investment. Sometimes called rebranding.

36 A 1980s planning measure revived in 2011 to arrest the decline of an area by providing incentives to invest in jobs and new economic activities.

37 Comprehensive Development Areas were programmes to demolished slum areas and re-house people in high-rise blocks (tower blocks). Expanded Towns were designed to take overspill from urban areas.

38 Quasi-retailing is a term used to describe activities in shopping centres such as building society offices and insurance brokers.

39 Discuss the merits of alternative explanations and conclude which is the most important.

40 A green belt is backed by law in the UK and surrounds the built up area. Wedges keep merging areas of built up land separate, whereas a corridor is similar but may not completely separate urban areas, only parts of them.

41 Because they preserve some of the rural character of an area even though there might be a club house or changing facilities.

42 The degree to which an area maintains rural character in its environment and economy.

43 'A transition town possesses a network of community-led responses to climate change and shrinking supplies of cheap energy, building resilience and happiness.' (Transition Network)

44 Opportunity deprivation takes the form of lack of access to education, health, work, social services and shops. Rural dwellers have the expense and difficulty of travelling longer distances for basic services, such as groceries, as well as paying more for them. This form of deprivation may be alleviated by online shopping (e.g. Tesco and Ocado) but these services depend on people having access to the net.